Robert Dick Wilson

Introductory Syriac Method and Manual

Robert Dick Wilson

Introductory Syriac Method and Manual

ISBN/EAN: 9783337246228

Printed in Europe, USA, Canada, Australia, Japan

Cover: Foto ©berggeist007 / pixelio.de

More available books at **www.hansebooks.com**

INTRODUCTORY

SYRIAC METHOD AND MANUAL

BY

ROBERT DICK WILSON, Ph.D.

PROFESSOR OF OLD TESTAMENT LANGUAGES AND HISTORY IN THE WESTERN THEOLOGICAL SEMINARY, ALLEGHENY, PA.

NEW YORK

CHARLES SCRIBNER'S SONS

1891

TO MY BELOVED PARENTS

THIS WORK

IS

RESPECTFULLY AND GRATEFULLY

DEDICATED

PREFACE.

THE plan of this METHOD AND MANUAL is in general the same as that of the corresponding "Introductory Hebrew Method and Manual" of Professor W. R. Harper, Ph.D. The following notes may be in place by way of explanation.

The first four chapters of Genesis (which are copied with variations from Nestle's "Syriac Grammar") are chosen because they afford the best means of comparison with the Hebrew of Professor Harper's "Manual."

The selections from the 10th to the 32d page, inclusive, lead up gradually from more easy to more difficult portions of the Peshito version. The last selection is the introductory portion of the history of Rabban Soma, possessed in manuscript by the author and never before published. Being printed in the Nestorian alphabet, it will be useful as an introduction to the East Syriac system of writing. For assistance in reading this selection the reader is referred especially to the note under Section I., Article 6, and to Article 6. 6. of the "Elements."

The "Notes and Observations" need no remark, except that the latter contain all of the main principles of Syriac grammar, while the former give all explanations necessary for a full understanding of the orthography, etymology, and syntax of the text.

The "Grammar Lessons" carry the student over all the articles of the "Elements of Syriac Grammar," with reviews of the same. The "Word Lessons" contain only such words as are not in the verses of Genesis, upon which the "Exercises" are largely based. When the grammar lesson has been upon a certain subject, the word lesson gives such words as throw light upon it; *e. g.*, in Lesson XI. the grammar lesson is on Lomadh Olaph verbs; the word lesson consists largely of Lomadh Olaph verbs. The vocabulary thus learned can be enlarged from the "Word Lists" on pp.

134–147. The "Exercises" are based upon the text of Genesis and upon the grammar and word lessons. They will be found, it is hoped, an excellent means of fixing in the memory the principles of grammar and the words of most common use. The "Exercises" can be supplemented by the transliteration of Genesis I., and by the literal translation of Genesis I.–IV., found at the end of the volume.

TABLE OF CONTENTS.

THE FIRST FOUR CHAPTERS OF GENESIS.

Chapter I.

1 ܒܪܫܝܬ ܒܪܐ ܐܠܗܐ ܝܬ ܫܡܝܐ ܘܝܬ ܐܪܥܐ. 2 ܘܐܪܥܐ ܗܘܬ ܬܘܗ
ܘܒܘܗ: ܘܚܫܘܟܐ ܥܠ ܐܦܝ ܬܗܘܡܐ: ܘܪܘܚܗ ܕܐܠܗܐ ܡܪܚܦܐ ܥܠ
ܐܦܝ ܡܝܐ. ܘܐܡܪ ܐܠܗܐ: ܢܗܘܐ ܢܘܗܪܐ: ܘܗܘܐ ܢܘܗܪܐ. 3 ܘܚܙܐ ܐܠܗܐ
ܠܢܘܗܪܐ ܕܫܦܝܪ. 4 ܘܦܪܫ ܐܠܗܐ ܒܝܢ ܢܘܗܪܐ ܠܚܫܘܟܐ. 5 ܘܩܪܐ ܐܠܗܐ
ܠܢܘܗܪܐ ܐܝܡܡܐ. ܘܠܚܫܘܟܐ ܩܪܐ ܠܠܝܐ: ܘܗܘܐ ܪܡܫܐ ܘܗܘܐ ܨܦܪܐ
ܝܘܡܐ ܚܕ. 6 ܘܐܡܪ ܐܠܗܐ ܢܗܘܐ ܪܩܝܥܐ ܒܡܨܥܬ ܡܝܐ. ܘܢܗܘܐ ܦܪܫ
ܒܝܢ ܡܝܐ ܠܡܝܐ. 7 ܘܥܒܕ ܐܠܗܐ ܐܪܩܝܥܐ: ܘܦܪܫ ܒܝܢ ܡܝܐ ܕܠܬܚܬ
ܡܢ ܐܪܩܝܥܐ: ܘܒܝܢ ܡܝܐ ܕܠܥܠ ܡܢ ܐܪܩܝܥܐ. ܘܗܘܐ ܗܟܢܐ. 8 ܘܩܪܐ
ܐܠܗܐ ܠܐܪܩܝܥܐ ܫܡܝܐ. ܘܗܘܐ ܪܡܫܐ ܘܗܘܐ ܨܦܪܐ ܝܘܡܐ ܕܬܪܝܢ. 9 ܘܐܡܪ
ܐܠܗܐ ܢܬܟܢܫܘܢ ܡܝܐ ܕܠܬܚܬ ܡܢ ܫܡܝܐ ܠܐܬܪܐ ܚܕ: ܘܬܬܚܙܐ ܝܒܝܫܬܐ:
ܘܗܘܐ ܗܟܢܐ. 10 ܘܩܪܐ ܐܠܗܐ ܠܝܒܝܫܬܐ ܐܪܥܐ: ܘܠܟܢܫܐ ܕܡܝܐ ܩܪܐ
ܝܡܡܐ. ܘܚܙܐ ܐܠܗܐ ܕܫܦܝܪ. 11 ܘܐܡܪ ܐܠܗܐ: ܬܦܩ ܐܪܥܐ ܬܕܐܐ ܥܣܒܐ
ܕܡܙܕܪܥ ܙܪܥܐ ܠܓܢܣܗ. ܘܐܝܠܢܐ ܕܦܐܪܐ ܕܥܒܕ ܦܐܪܐ ܠܓܢܣܗ: ܕܙܪܥܗ ܒܗ
ܥܠ ܐܪܥܐ: ܘܗܘܐ ܗܟܢܐ. 12 ܘܐܦܩܬ ܐܪܥܐ ܬܕܐܐ ܥܣܒܐ ܕܡܙܕܪܥ ܙܪܥܐ
ܠܓܢܣܗ: ܘܐܝܠܢܐ ܕܥܒܕ ܦܐܪܐ ܕܙܪܥܗ ܒܗ ܠܓܢܣܗ. ܘܚܙܐ ܐܠܗܐ

ܕܫܦܝܪ. 13 ܘܗܘܐ ܪܡܫܐ ܘܗܘܐ ܨܦܪܐ. ܝܘܡܐ ܕܬܠܬܐ܀ 14 ܘܐܡܪ ܐܠܗܐ
ܢܗܘܘܢ ܢܗܝܪܐ ܒܐܪܩܝܥܐ ܕܫܡܝܐ. ܠܡܦܪܫܘ ܒܝܬ ܐܝܡܡܐ ܠܠܠܝܐ:
ܘܢܗܘܘܢ ܠܐܬܘܬܐ ܘܠܙܒܢܐ ܘܠܝܘܡܬܐ ܘܠܫܢܝܐ. 15 ܘܢܗܘܘܢ ܡܢܗܪܝܢ
ܒܐܪܩܝܥܐ ܕܫܡܝܐ ܠܡܢܗܪܘ ܥܠ ܐܪܥܐ: ܘܗܘܐ ܗܟܢܐ. 16 ܘܥܒܕ ܐܠܗܐ ܬܪܝܢ
ܢܗܝܪܐ ܪܘܪܒܐ: ܢܗܝܪܐ ܪܒܐ ܠܫܘܠܛܢܐ ܕܐܝܡܡܐ: ܘܢܗܝܪܐ ܙܥܘܪܐ
ܠܫܘܠܛܢܐ ܕܠܠܝܐ: ܘܟܘܟܒܐ. 17 ܘܝܗܒ ܐܢܘܢ ܐܠܗܐ ܒܐܪܩܝܥܐ ܕܫܡܝܐ.
ܠܡܢܗܪܘ ܥܠ ܐܪܥܐ. 18 ܘܠܡܫܠܛܘ ܒܐܝܡܡܐ ܘܒܠܠܝܐ: ܘܠܡܦܪܫܘ
ܒܝܬ ܢܘܗܪܐ ܠܚܫܘܟܐ: ܘܚܙܐ ܐܠܗܐ ܕܫܦܝܪ. 19 ܘܗܘܐ ܪܡܫܐ ܘܗܘܐ ܨܦܪܐ
ܝܘܡܐ ܕܐܪܒܥܐ܀ 20 ܘܐܡܪ ܐܠܗܐ ܢܪܚܫܘܢ ܡܝܐ ܪܚܫܐ ܢܦܫܐ ܚܝܬܐ ܘܦܪܚܬܐ
ܬܦܪܚ ܥܠ ܐܪܥܐ ܥܠ ܐܦܝ ܐܪܩܝܥܐ ܕܫܡܝܐ. 21 ܘܒܪܐ ܐܠܗܐ ܬܢܝܢܐ ܪܘܪܒܐ
ܘܟܠ ܢܦܫܐ ܚܝܬܐ ܕܪܚܫܐ ܕܐܪܚܫܘ ܡܝܐ ܠܓܢܣܗܘܢ. ܘܟܠ ܦܪܚܬܐ ܕܓܦܐ
ܠܓܢܣܗ: ܘܚܙܐ ܐܠܗܐ ܕܫܦܝܪ. 22 ܘܒܪܟ ܐܢܘܢ ܐܠܗܐ ܘܐܡܪ ܠܗܘܢ: ܦܘܫܘ
ܘܣܓܘ: ܘܡܠܘ ܡܝܐ ܕܒܝܡܡܐ: ܘܦܪܚܬܐ ܬܣܓܐ ܒܐܪܥܐ. 23 ܘܗܘܐ ܪܡܫܐ
ܘܗܘܐ ܨܦܪܐ. ܝܘܡܐ ܕܚܡܫܐ܀ 24 ܘܐܡܪ ܐܠܗܐ. ܬܦܩ ܐܪܥܐ ܢܦܫܐ ܚܝܬܐ
ܠܓܢܣܗ: ܒܥܝܪܐ ܘܪܚܫܐ: ܘܚܝܘܬܐ ܕܐܪܥܐ ܠܓܢܣܗ: ܘܗܘܐ ܗܟܢܐ.
25 ܘܥܒܕ ܐܠܗܐ ܚܝܘܬܐ ܕܐܪܥܐ ܠܓܢܣܗ: ܘܒܥܝܪܐ ܠܓܢܣܗ: ܘܟܠܗ ܪܚܫܐ
ܕܐܪܥܐ ܠܓܢܣܗܘܢ: ܘܚܙܐ ܐܠܗܐ ܕܫܦܝܪ. 26 ܘܐܡܪ ܐܠܗܐ: ܢܥܒܕ ܐܢܫܐ
ܒܨܠܡܢ ܐܝܟ ܕܡܘܬܢ. ܘܢܫܬܠܛܘܢ ܒܢܘܢܝ ܝܡܐ: ܘܒܦܪܚܬܐ ܕܫܡܝܐ.
ܘܒܒܥܝܪܐ ܘܒܟܠܗ ܚܝܘܬܐ ܕܐܪܥܐ. ܘܒܟܠܗ ܪܚܫܐ ܕܪܚܫ ܥܠ ܐܪܥܐ.

[27] ܘܒܪܐ ܐܠܗܐ ܠܐܕܡ ܒܨܠܡܗ. ܒܨܠܡܐ ܕܐܠܗܐ ܒܪܝܗܝ. ܕܟܪܐ ܘܢܩܒܬܐ ܒܪܐ ܐܢܘܢ. [28] ܘܒܪܟ ܐܢܘܢ ܐܠܗܐ: ܘܐܡܪ ܠܗܘܢ ܐܠܗܐ: ܦܪܘ ܘܣܓܘ: ܘܡܠܘܗ ܐܪܥܐ ܘܐܬܟܒܫܘܗ. ܘܫܠܛܘ ܒܢܘܢܝ ܝܡܐ: ܘܒܦܪܚܬܐ ܕܫܡܝܐ: ܘܒܒܥܝܪܐ ܘܒܟܠܗ ܚܝܘܬܐ ܕܪܚܫܐ ܥܠ ܐܪܥܐ. [29] ܘܐܡܪ ܐܠܗܐ: ܗܐ ܝܗܒܬ ܠܟܘܢ ܟܠܗ ܥܣܒܐ ܕܙܪܥܐ ܙܪܥܗ ܕܥܠ ܐܦܝ ܟܠܗ ܐܪܥܐ. ܘܟܠ ܐܝܠܢ ܕܐܝܬ ܒܗ ܦܐܪܐ ܐܝܠܢܐ ܕܙܪܥܗ ܡܙܕܪܥ: ܠܟܘܢ ܢܗܘܐ ܠܡܐܟܘܠܬܐ. ܘܠܟܠܗ ܚܝܘܬܐ ܕܒܪܐ: [30] ܘܠܟܠܗ ܦܪܚܬܐ ܕܫܡܝܐ: ܘܠܟܠ ܕܪܚܫ ܥܠ ܐܪܥܐ ܕܐܝܬ ܒܗ ܢܦܫܐ ܚܝܬܐ: ܘܟܠܗ ܝܘܪܩܐ ܕܥܣܒܐ ܠܡܐܟܘܠܬܐ: ܘܗܘܐ ܗܟܢܐ. [31] ܘܚܙܐ ܐܠܗܐ ܟܠ ܕܥܒܕ: ܘܗܐ ܛܒ ܫܦܝܪ. ܘܗܘܐ ܪܡܫܐ ܘܗܘܐ ܨܦܪܐ ܝܘܡܐ ܕܫܬܐ.

Chapter II.

[1] ܘܐܫܬܠܡܘ ܫܡܝܐ ܘܐܪܥܐ: ܘܟܠܗ ܚܝܠܗܘܢ. [2] ܘܫܠܡ ܐܠܗܐ ܒܝܘܡܐ ܫܒܝܥܝܐ ܥܒܕܘܗܝ ܕܥܒܕ. ܘܐܬܬܢܝܚ ܒܝܘܡܐ ܫܒܝܥܝܐ ܡܢ ܟܠܗܘܢ ܥܒܕܘܗܝ ܕܥܒܕ. [3] ܘܒܪܟ ܐܠܗܐ ܠܝܘܡܐ ܫܒܝܥܝܐ ܘܩܕܫܗ. ܡܛܠ ܕܒܗ ܐܬܬܢܝܚ ܡܢ ܟܠܗܘܢ ܥܒܕܘܗܝ ܕܒܪܐ ܐܠܗܐ ܠܡܥܒܕ. [4] ܗܠܝܢ ܬܘܠܕܬܐ ܕܫܡܝܐ ܘܕܐܪܥܐ ܟܕ ܐܬܒܪܝܘ: ܒܝܘܡܐ ܕܥܒܕ ܡܪܝܐ ܐܠܗܐ ܫܡܝܐ ܘܐܪܥܐ. [5] ܘܟܠܗܘܢ ܐܝܠܢܐ ܕܚܩܠܐ ܥܕܠܐ ܢܗܘܘܢ ܒܐܪܥܐ. ܘܟܠܗ ܥܣܒܐ ܕܚܩܠܐ ܥܕܠܐ ܐܢܒܐ: ܡܛܠ ܕܠܐ ܐܚܬ ܡܪܝܐ ܐܠܗܐ ܡܛܪܐ ܥܠ [ܐܦܝ] ܐܪܥܐ. ܘܐܢܫ ܠܝܬ ܗܘܐ ܕܢܦܠܘܚ ܒܐܪܥܐ. [6] ܘܡܥܝܢܐ ܣܠܩ ܗܘܐ ܡܢ ܐܪܥܐ: ܘܡܫܩܐ

ܗܘܳܐ [ܠ]ܟܽܠܶܗ ܐܰܦܰܝ ܐܰܪܥܳܐ. ⁷ ܘܰܓܒܰܠ ܡܳܪܝܳܐ ܐܰܠܳܗܳܐ ܠܐܳܕܳܡ ܥܰܦܪܳܐ ܡܶܢ ܐܰܕܡܬܳܐ: ܘܢܰܦܰܚ ܒܰܐܦܰܘܗܝ ܢܶܫܡܬܳܐ ܕܚܰܝܶܐ. ܘܰܗܘܳܐ ܐܳܕܳܡ ܠܢܰܦܫܳܐ ܚܰܝܬܳܐ. ⁸ ܘܰܢܨܰܒ ܡܳܪܝܳܐ ܐܰܠܳܗܳܐ ܦܰܪܕܰܝܣܳܐ ܒܥܶܕܶܢ ܡܶܢ ܩܽܕܡ: ܘܣܳܡ ܬܰܡܳܢ ܠܐܳܕܳܡ ܕܓܒܰܠ. ⁹ ܘܰܐܰܘܥܺܝ ܡܳܪܝܳܐ ܐܰܠܳܗܳܐ ܡܶܢ ܐܰܪܥܳܐ ܟܽܠ ܐܺܝܠܳܢ ܕܰܪܓܺܝܓ ܠܚܶܙܘܳܐ: ܘܫܰܦܺܝܪ ܠܡܶܐܟܰܠ: ܘܐܺܝܠܳܢܳܐ ܕܚܰܝܶܐ ܒܡܶܨܥܰܬ ܦܰܪܕܰܝܣܳܐ: ܘܐܺܝܠܳܢܳܐ ܕܺܝܳܕܥܳܐ ܕܛܳܒܬܳܐ ܘܰܒܺܝܫܬܳܐ. ¹⁰ ܘܢܰܗܪܳܐ ܢܳܦܶܩ ܗܘܳܐ ܡܶܢ ܥܶܕܶܢ ܠܰܡܫܰܩܳܝܽܘ ܠܦܰܪܕܰܝܣܳܐ: ܘܡܶܢ ܬܰܡܳܢ ܦܳܪܶܫ ܗܘܳܐ ܠܰܐܪܒܥܳܐ ܪ̈ܺܫܺܝܢ. ¹¹ ܫܡܶܗ ܕܚܰܕ ܦܺܝܫܽܘܢ: ܗܽܘ ܕܚܳܕܰܪ ܠܟܽܠܳܗ ܐܰܪܥܳܐ ܕܚܰܘܺܝܠܳܐ: ܕܬܰܡܳܢ ܐܺܝܬ ܕܰܗܒܳܐ. ¹² ܘܕܰܗܒܳܗ ܕܰܐܪܥܳܐ ܗܳܝ ܛܳܒ: ܬܰܡܳܢ ܐܺܝܬ ܒܶܕܽܘܠܚܳܐ ܘܟܺܐܦܳܐ ܕܒܽܘܪܽܘܠܳܐ. ¹³ ܘܫܡܶܗ ܕܢܰܗܪܳܐ ܕܬܰܪܝܳܢ ܓܺܝܚܳܘܢ: ܗܽܘ ܕܚܳܕܰܪ ܠܟܽܠܳܗ ܐܰܪܥܳܐ ܕܟܽܘܫ. ¹⁴ ܘܫܡܶܗ ܕܢܰܗܪܳܐ ܕܬܠܳܬܳܐ ܕܶܩܠܰܬ: ܗܽܘ ܕܳܐܙܶܠ ܠܡܰܕܢܚܳܐ ܕܰܐܬܽܘܪ. ܘܢܰܗܪܳܐ ܕܰܐܪܒܥܳܐ ܗܽܘ ܦܪܳܬ. ¹⁵ ܘܕܰܒܪܶܗ ܡܳܪܝܳܐ ܐܰܠܳܗܳܐ ܠܐܳܕܳܡ ܘܰܐܫܪܝܶܗ ܒܦܰܪܕܰܝܣܳܐ ܕܥܶܕܶܢ. ܕܢܶܦܠܚܺܝܘܗܝ ܘܢܶܛܪܺܝܘܗܝ. ¹⁶ ܘܦܰܩܶܕ ܡܳܪܝܳܐ ܐܰܠܳܗܳܐ ܠܐܳܕܳܡ ܘܶܐܡܰܪ ܠܶܗ: ܡܶܢ ܟܽܠܗܽܘܢ ܐܺܝܠܳܢܶܐ ܕܒܦܰܪܕܰܝܣܳܐ ܡܶܐܟܰܠ ܬܺܐܟܽܘܠ: ¹⁷ ܘܡܶܢ ܐܺܝܠܳܢܳܐ ܕܝܳܕܥܺܝܢ ܛܳܒܬܳܐ ܘܒܺܝܫܬܳܐ ܠܳܐ ܬܺܐܟܽܘܠ ܡܶܢܶܗ. ܡܶܛܽܠ ܕܒܝܰܘܡܳܐ ܕܬܺܐܟܽܘܠ ܡܶܢܶܗ ܡܳܐܬ ܬܡܽܘܬ. ¹⁸ ܘܶܐܡܰܪ ܡܳܪܝܳܐ ܐܰܠܳܗܳܐ: ܠܳܐ ܫܰܦܺܝܪ ܕܢܶܗܘܶܐ ܐܳܕܳܡ ܒܰܠܚܽܘܕܰܘܗܝ. ܐܶܥܒܶܕ ܠܶܗ ܡܰܥܕܪܳܢܳܐ ܐܰܟܘܳܬܶܗ. ¹⁹ ܘܰܓܒܰܠ ܡܳܪܝܳܐ ܐܰܠܳܗܳܐ ܡܶܢ ܐܰܪܥܳܐ ܟܽܠܗ ܚܰܝܘܳܬܳܐ ܕܒܰܪܳܐ: ܘܟܽܠܗ ܦܳܪܰܚܬܳܐ ܕܰܫܡܰܝܳܐ: ܘܰܐܝܬܺܝ ܐܶܢܽܘܢ ܠܘܳܬ ܐܳܕܳܡ ܕܢܶܚܙܶܐ ܡܳܢܳܐ ܩܳܪܶܐ ܠܗܽܘܢ: ܘܟܽܠ ܕܰܩܪܳܐ ܠܶܗ ܐܳܕܳܡ ܢܰܦܫܳܐ ܚܰܝܬܳܐ ܗܽܘ ܗ̄ܘ ܫܡܶܗ. ²⁰ ܘܰܩܪܳܐ ܐܳܕܳܡ ܫܡܳܗܶܐ ܠܟܽܠܗ ܒܥܺܝܪܳܐ: ܘܰܠܟܽܠܗ ܦܳܪܰܚܬܳܐ ܕܰܫܡܰܝܳܐ:

ܘܠܟܠܗ ܚܝܘܬܐ ܕܒܪܐ. ܘܠܐܕܡ ܠܐ ܐܫܬܟܚ [ܠܗ] ܡܥܕܪܢܐ ܐܟܘܬܗ.
²¹ ܘܐܪܡܝ ܡܪܝܐ ܐܠܗܐ ܫܢܬܐ ܥܠ ܐܕܡ ܘܕܡܟ: ܘܢܣܒ ܚܕܐ ܡܢ
ܥܠܥܘܗܝ: ܘܐܚܠܡ ܒܣܪܐ ܚܠܦܝܗ. ²² ܘܒܢܐ ܡܪܝܐ ܐܠܗܐ ܠܥܠܥܐ ܕܢܣܒ
ܡܢ ܐܕܡ ܠܐܢܬܬܐ ܘܐܝܬܝܗ ܠܐܕܡ. ²³ ܘܐܡܪ ܐܕܡ. ܗܕܐ ܐܝܬܝܗ ܓܪܡܐ ܡܢ
ܓܪܡܝ: ܘܒܣܪܐ ܡܢ ܒܣܪܝ: ܗܕܐ ܬܬܩܪܐ ܐܢܬܬܐ ܡܛܠ ܕܡܢ ܓܒܪܐ
ܢܣܝܒܐ. ²⁴ ܡܛܠ ܗܢܐ ܢܫܒܘܩ ܓܒܪܐ ܠܐܒܘܗܝ ܘܠܐܡܗ: ܘܢܩܦ
ܠܐܢܬܬܗ. ܘܢܗܘܘܢ ܬܪܝܗܘܢ ܚܕ ܒܣܪ. ²⁵ ܘܗܘܘ ܬܪܝܗܘܢ ܥܪܛܠܝܢ: ܐܕܡ
ܘܐܢܬܬܗ ܘܠܐ ܒܗܬܝܢ܀

Chapter III.

¹ ܘܚܘܝܐ ܥܪܝܡ ܗܘܐ ܡܢ ܟܠܗ ܚܝܘܬܐ ܕܒܪܐ ܕܥܒܕ ܡܪܝܐ ܐܠܗܐ.
ܘܐܡܪ ܚܘܝܐ ܠܐܢܬܬܐ. ܫܪܝܪܐܝܬ ܐܡܪ ܐܠܗܐ ܕܠܐ ܬܐܟܠܘܢ ܡܢ ܟܠ ܐܝܠܢܝ
ܦܪܕܝܣܐ. ² ܘܐܡܪܬ ܐܢܬܬܐ ܠܚܘܝܐ. ܕܡܢ ܦܐܪܐ ܐܝܠܢܐ ܕܦܪܕܝܣܐ ܡܟܠܗܘܢ
ܬܐܟܘܠ. ³ ܘܡܢ ܦܐܪܐ ܐܝܠܢܐ ܕܒܡܨܥܬܗ ܕܦܪܕܝܣܐ ܐܡܪ ܠܢ ܐܠܗܐ ܕܠܐ ܬܐܟܠܘܢ
ܡܢܗ ܘܠܐ ܬܬܩܪܒܘܢ ܠܗ ܕܠܐ ܬܡܘܬܘܢ. ⁴ ܘܐܡܪ ܚܘܝܐ ܠܐܢܬܬܐ ܠܐ ܡܡܬ
ܬܡܘܬܘܢ. ⁵ ܡܛܠ ܕܝܕܥ ܠܗ ܐܠܗܐ. ܕܒܝܘܡܐ ܕܐܟܠܝܢ ܐܢܬܘܢ ܡܢܗ.
ܡܬܦܬܚܢ ܥܝܢܝܟܘܢ ܘܗܘܝܢ ܐܢܬܘܢ ܐܝܟ ܐܠܗܐ ܝܕܥܝ ܛܒܬܐ ܘܒܝܫܬܐ. ⁶ ܘܚܙܬ
ܐܢܬܬܐ ܕܫܦܝܪ ܐܝܠܢܐ ܠܡܐܟܘܠܐ ܘܪܓܝܓ ܗܘ ܠܥܝܢܐ ܘܡܪܓܓ ܐܝܠܢܐ ܠܡܣܬܟܠܘ
ܒܗ. ܘܢܣܒܬ ܡܢ ܦܐܪܘܗܝ ܘܐܟܠܬ. ܘܝܗܒܬ ܐܦ ܠܒܥܠܗ ܥܡܗ

ܘܐܟܠ. ⁷ ܘܐܬܦܬܚ[ܝ] ܥܝܢܝ ܬܪܝܗܘܢ. ܘܝܕܥܘ ܕܥܪܛܠܝܐ ܐܢܘܢ. ܘܚܝܛܘ
ܛܪܦܐ ܕܬܐܢܐ. ܘܥܒܕܘ ܠܗܘܢ ܙܪܙܐ. ⁸ ܘܫܡܥܘ ܩܠܗ ܕܡܪܝܐ ܐܠܗܐ
ܡܗܠܟ ܒܦܪܕܝܣܐ ܠܡܦܢܝܗ ܕܝܘܡܐ. ܘܐܛܫܝ[ܘ] ܐܕܡ ܘܐܢܬܬܗ ܡܢ ܩܕܡ
ܡܪܝܐ ܐܠܗܐ ܒܝܬ ܐܝܠܢܐ ܕܦܪܕܝܣܐ. ⁹ ܘܩܪܐ ܡܪܝܐ ܐܠܗܐ ܠܐܕܡ ܘܐܡܪ ܠܗ.
ܐܝܟܐ ܐܢܬ ܐܕܡ. ¹⁰ ܘܐܡܪ ܩܠܟ ܫܡܥܬ ܒܦܪܕܝܣܐ ܘܕܚܠܬ ܕܥܪܛܠܝ ܐܢܐ
ܘܐܬܛܫܝܬ. ¹¹ ܘܐܡܪ ܠܗ ܡܪܝܐ ܡܢܘ ܚܘܝܟ ܕܥܪܛܠܝ ܐܢܬ. ܗܐ ܡܢ
ܐܝܠܢܐ ܕܦܩܕܬܟ ܕܠܐ ܬܐܟܘܠ ܡܢܗ ܐܟܠܬ. ¹² ܘܐܡܪ ܐܕܡ. ܐܢܬܬܐ ܕܝܗܒܬ
ܥܡܝ ܗܝ ܝܗܒܬ ܠܝ ܡܢ ܐܝܠܢܐ ܘܐܟܠܬ. ܘܐܡܪ ܡܪܝܐ ܐܠܗܐ ܠܐܢܬܬܐ.
¹³ ܡܢܐ ܗܕܐ ܥܒܕܬܝ. ܘܐܡܪܬ ܐܢܬܬܐ. ܚܘܝܐ ܐܛܥܝܢܝ ܘܐܟܠܬ. ¹⁴ ܘܐܡܪ
ܡܪܝܐ ܐܠܗܐ ܠܚܘܝܐ. ܕܗܟܢܐ ܥܒܕܬ ܗܕܐ. ܠܝܛ ܐܢܬ ܡܢ ܟܠܗ ܒܥܝܪܐ
ܘܡܢ ܟܠܗ ܚܝܘܬܐ ܕܒܪܐ. ܘܥܠ ܚܕܝܟ ܬܗܠܟ. ܘܥܦܪܐ ܬܐܟܘܠ ܟܠ
ܝܘܡܝ ܚܝܝܟ. ¹⁵ ܘܒܥܠܕܒܒܘܬܐ ܐܣܝܡ ܒܝܢܝܟ ܠܐܢܬܬܐ. ܘܒܝܢ ܙܪܥܟ
ܠܙܪܥܗ. ܗܘ ܢܬܘܫ ܪܫܟ ܘܐܢܬ ܬܬܘܫܝܘܗܝ ܥܩܒܗ. ¹⁶ ܘܠܐܢܬܬܐ ܐܡܪ
ܡܣܓܐ ܐܣܓܐ ܟܐܒܝܟܝ ܘܒܛܢܟܝ. ܘܒܟܐܒܐ ܬܐܠܕܝܢ ܒܢܝܐ. ܘܠܘܬ
ܒܥܠܟܝ ܬܬܦܢܝܢ. ܘܗܘ ܢܫܠܛ ܒܟܝ. ¹⁷ ܘܠܐܕܡ ܐܡܪ ܕܠܐ ܫܡܥܬ
ܠܩܠܗ ܕܐܢܬܬܟ. ܘܐܟܠܬ ܡܢ ܐܝܠܢܐ ܕܦܩܕܬܟ ܘܐܡܪܬ ܠܟ ܕܠܐ ܬܐܟܘܠ
ܡܢܗ. ܠܝܛܐ ܐܪܥܐ ܡܛܠܬܟ ܒܥܡܠܐ ܬܐܟܠܝܗ ܟܠ ܝܘܡܝ ܚܝܝܟ.
¹⁸ ܟܘܒܐ ܘܕܪܕܪܐ ܬܦܩ ܠܟ. ܘܬܐܟܘܠ ܥܣܒܐ ܕܚܩܠܐ. ¹⁹ ܒܕܘܥܬܐ ܕܐܦܝܟ
ܬܐܟܘܠ ܠܚܡܟ. ܥܕܡܐ ܕܬܗܦܘܟ ܠܐܪܥܐ ܕܡܢܗ ܐܬܢܣܒܬ. ܡܛܠ ܕܥܦܪܐ ܐܢܬ

ܘܠܥܦܪܐ ܬܗܦܘܟ. [20] ܘܩܪܐ ܐܕܡ ܫܡܗ ܕܐܢܬܬܗ ܚܘܐ. ܡܛܠ ܕܗܝ ܗܘܬ
ܐܡܐ ܕܟܠ ܕܚܝ. [21] ܘܥܒܕ ܡܪܝܐ ܐܠܗܐ ܠܐܕܡ ܘܠܐܢܬܬܗ ܟܘܬܝܢܝܬܐ ܕܡܫܟܐ
ܘܐܠܒܫ ܐܢܘܢ. [22] ܘܐܡܪ ܡܪܝܐ ܐܠܗܐ. ܗܐ ܐܕܡ ܗܘܐ ܐܝܟ ܚܕ ܡܢܢ.
ܠܡܕܥ ܛܒܬܐ ܘܒܝܫܬܐ. ܘܗܫܐ ܕܠܡܐ ܢܘܫܛ ܐܝܕܗ ܘܢܤܒ ܐܦ ܡܢ
ܐܝܠܢܐ ܕܚܝܐ. ܘܢܐܟܘܠ ܘܢܚܐ ܠܥܠܡ. [23] ܘܫܕܪܗ ܡܪܝܐ ܐܠܗܐ ܡܢ
ܦܪܕܝܤܐ ܕܥܕܢ. ܠܡܦܠܚ ܐܪܥܐ ܕܐܬܢܤܒ ܡܢ ܬܡܢ. [24] ܘܐܦܩܗ ܡܪܝܐ
ܠܐܕܡ ܘܐܫܪܝ ܡܢ ܡܕܢܚܐ ܠܦܪܕܝܤܐ ܕܥܕܢ ܟܪܘܒܐ ܘܫܢܢܐ ܕܤܝܦܐ
ܕܡܬܗܦܟܐ ܠܡܛܪ ܐܘܪܚܐ ܕܐܝܠܢܐ ܕܚܝܐ.

Chapter IV.

[1] ܘܐܕܡ ܝܕܥ ܠܚܘܐ ܐܢܬܬܗ. ܘܒܛܢܬ ܘܝܠܕܬ ܠܩܐܝܢ. ܘܐܡܪܬ. ܩܢܝܬ
ܓܒܪܐ ܠܡܪܝܐ. [2] ܘܐܘܤܦܬ ܠܡܐܠܕ ܠܐܚܘܗܝ ܠܗܒܝܠ. ܘܗܘܐ ܗܒܝܠ ܪܥܐ
ܥܢܐ. ܘܩܐܝܢ ܗܘܐ ܦܠܚ ܒܐܪܥܐ. [3] ܘܗܘܐ ܡܢ ܒܬܪ ܝܘܡܬܐ. ܘܐܝܬܝ
ܩܐܝܢ ܡܢ ܦܐܪܐ ܕܐܪܥܐ. ܩܘܪܒܢܐ ܠܡܪܝܐ. [4] ܘܗܒܝܠ ܐܝܬܝ ܐܦ ܗܘ ܡܢ
ܒܘܟܪܐ ܕܥܢܗ ܘܡܢ ܫܡܝܢܝܗܝܢ. ܘܐܬܪܥܝ ܡܪܝܐ ܒܗܒܝܠ ܘܒܩܘܪܒܢܗ.
[5] ܘܒܩܐܝܢ ܘܒܩܘܪܒܢܗ ܠܐ ܐܬܪܥܝ. ܘܐܬܥܩܬ ܠܩܐܝܢ ܛܒ. ܘܐܬܟܡܪܘ
ܐܦܘܗܝ. [6] ܘܐܡܪ ܡܪܝܐ ܠܩܐܝܢ. ܠܡܢܐ ܐܬܥܩܬ ܠܟ. ܘܠܡܢܐ ܐܬܟܡܪܘ
ܐܦܝܟ. [7] ܗܐ ܐܢ ܬܛܐܒ ܡܩܒܠ. ܘܐܢ ܠܐ ܬܛܐܒ. ܥܠ ܬܪܥܐ ܚܛܝܬܐ ܪܒܥܐ.

ܐܠܝܟ ܬܬܦܢܐ ܪܓܬܗ ܘܗܘ ܬܫܬܠܛ ܒܗ. 8 ܘܐܡܪ ܩܐܝܢ ܠܗܒܝܠ
ܐܚܘܗܝ. ܢܪܕܐ ܠܦܩܥܬܐ. ܘܗܘܐ ܕܟܕ ܗܘܘ ܒܦܩܥܬܐ. ܩܡ ܩܐܝܢ ܥܠ ܗܒܝܠ
ܐܚܘܗܝ ܘܩܛܠܗ. 9 ܘܐܡܪ ܡܪܝܐ ܠܩܐܝܢ. ܐܝܟܐ ܗܘ ܗܒܝܠ ܐܚܘܟ. ܘܐܡܪ
ܠܐ ܝܕܥ ܐܢܐ. ܢܛܘܪܗ ܐܢܐ ܕܐܚܝ. 10 ܘܐܡܪ. ܡܢܐ ܥܒܕܬ. ܩܠܐ ܕܕܡܗ
ܕܐܚܘܟ ܩܥܐ ܠܘܬܝ ܡܢ ܐܪܥܐ. 11 ܘܗܫܐ ܠܝܛ ܐܢܬ ܡܢ ܐܪܥܐ. ܕܦܬܚܬ
ܦܘܡܗ ܘܩܒܠܬ ܕܡܗ ܕܐܚܘܟ ܡܢ ܐܝܕܝܟ. 12 ܘܡܐ ܕܬܦܠܘܚ ܒܐܪܥܐ ܠܐ
ܬܘܣܦ ܕܬܬܠ ܠܟ ܚܝܠܗ. ܙܐܥ ܘܢܐܕ ܬܗܘܐ ܒܐܪܥܐ. 13 ܘܐܡܪ ܩܐܝܢ
ܠܡܪܝܐ. ܪܒܐ ܗܝ ܣܟܠܘܬܝ ܡܢ ܕܢܫܬܒܩ. 14 ܗܐ ܐܦܩܬܢܝ ܝܘܡܢܐ ܡܢ
ܐܦܝ ܐܪܥܐ. ܘܡܢ ܩܕܡܝܟ ܐܛܫܐ. ܘܐܗܘܐ ܙܐܥ ܘܢܐܕ ܒܐܪܥܐ. ܘܟܠ ܡܢ
ܕܢܫܟܚܢܝ ܢܩܛܠܢܝ. 15 ܘܐܡܪ ܠܗ ܡܪܝܐ. ܠܐ ܗܟܢܐ ܟܠ ܩܛܘܠܐ. ܩܐܝܢ
ܫܒܥܐ ܥܘܢܕܢܝܢ ܢܬܦܪܥ. ܘܣܡ ܡܪܝܐ ܐܬܐ ܥܠ ܩܐܝܢ ܕܠܐ ܢܩܛܠܝܘܗܝ ܟܠ ܡܢ
ܕܢܫܟܚܝܘܗܝ ܠܗ. 16 ܘܢܦܩ ܩܐܝܢ ܡܢ ܩܕܡ ܡܪܝܐ. ܘܝܬܒ ܒܐܪܥܐ ܕܢܘܕ
ܡܢ ܡܕܢܚܗ ܕܥܕܢ. 17 ܘܝܕܥ ܩܐܝܢ ܠܐܢܬܬܗ. ܘܒܛܢܬ ܘܝܠܕܬ ܠܚܢܘܟ.
ܘܗܘܐ ܒܢܐ ܩܪܝܬܐ. ܘܩܪܐ ܫܡܗ ܕܩܪܝܬܐ ܥܠ ܫܡ ܒܪܗ ܚܢܘܟ.
18 ܘܐܬܝܠܕ ܠܚܢܘܟ ܥܝܪܕ. ܘܥܝܪܕ ܐܘܠܕ ܠܡܚܘܝܐܝܠ. ܘܡܚܘܝܐܝܠ ܐܘܠܕ
ܠܡܬܘܫܐܝܠ. ܘܡܬܘܫܐܝܠ ܐܘܠܕ ܠܠܡܟ. 19 ܘܠܡܟ ܢܣܒ ܠܗ ܬܪܬܝܢ
ܢܫܝܢ. ܫܡܗ ܕܚܕܐ ܥܕܐ ܘܫܡܗ ܕܐܚܪܬܐ ܨܠܐ. 20 ܘܝܠܕܬ ܥܕܐ
ܠܝܒܠ. ܗܘ ܗܘܐ ܐܒܐ ܠܥܡܘܪܝ ܡܫܟܢܐ ܘܩܢܝܢܐ. 21 ܘܫܡܐ
ܕܐܚܘܗܝ ܝܘܒܠ. ܗܘ ܗܘܐ ܐܒܐ ܠܟܠ ܕܐܚܕ ܟܢܪܐ ܘܩܝܬܪܐ. 22 ܘܨܠܐ ܐܦ

ܗ̈ܘ ܡܠܦ ܠܟܠܗܘܢ. ܐܘܡܢܝ ܟܠ ܥܒܕ ܕܢܚܫܐ ܘܕܦܪܙܠܐ. ܘܚܬܗ
ܕܬܘܒܠܩܝܢ ܢܥܡܐ. 23 ܘܐܡܪ ܠܡܟ ܠܢ̈ܫܘܗܝ. ܥܕܐ ܘܨܠܐ ܫܡܥܝܢ ܩܠܝ
ܢ̈ܫܐ ܕܠܡܟ ܨܘܬܝܢ ܠܡܐܡܪܝ. ܡܛܠ ܕܓܒܪܐ ܩܛܠܬ ܠܡܚܘܬܝ ܘܛܠܝܐ
ܒܡܘܡܫܝ. 24 ܡܛܠ ܕܫܒܥܐ ܢܬܦܪܥ ܩܐܝܢ. ܘܠܡܟ ܫܒܥܝܢ
ܘܫܒܥܐ. 25 ܘܝܕܥ ܐܕܡ ܬܘܒ ܠܚܘܐ ܐܢܬܬܗ. ܘܒܛܢܬ ܘܝܠܕܬ ܒܪܐ.
ܘܩܪܬ ܫܡܗ ܫܝܬ. ܡܛܠ ܕܐܩܝܡ ܠܝ ܐܠܗܐ ܙܪܥܐ ܐܚܪܢܐ ܚܠܦ ܗܒܝܠ
ܕܩܛܠܗ ܩܐܝܢ. 26 ܘܠܫܝܬ ܐܦ ܠܗ. ܐܬܝܠܕ ܠܗ ܒܪܐ. ܘܩܪܐ ܫܡܗ
ܐܢܘܫ. ܗܝܕܝܢ ܫܪܝ ܠܡܩܪܐ ܒܫܡܗ ܕܡܪܝܐ.

PSALM II.

---

¹ ܠܡܢܐ ܪܓܫܘ ܥܡܡܐ : ܘܐܡܘܬܐ ܪܢܝ ܣܪܝܩܬܐ ܀ ² ܩܡܘ ܡܠܟܐ ܕܐܪܥܐ
ܘܫܠܝܛܢܐ : ܘܐܬܡܠܟܘ ܐܟܚܕܐ : ܥܠ ܡܪܝܐ ܘܥܠ ܡܫܝܚܗ ܀ ³ ܢܦܣܘܩ
ܫܘܫܠܬܗܘܢ : ܘܢܫܕܐ ܡܢܢ ܢܝܪܗܘܢ ܀ ⁴ ܕܝܬܒ ܒܫܡܝܐ ܢܓܚܟ : ܘܡܪܝܐ
ܢܡܝܩ ܒܗܘܢ ܀ ⁵ ܗܝܕܝܢ ܢܡܠܠ ܥܡܗܘܢ ܒܪܘܓܙܗ : ܘܒܚܡܬܗ ܢܕܠܚ
ܐܢܘܢ ܀ ⁶ ܐܢܐ ܐܩܝܡܬܗ ܡܠܟܐ ܥܠ ܨܗܝܘܢ ܛܘܪܐ ܕܩܘܕܫܗ ܘܠܡܫܩܠ ܥܠ
ܦܘܩܕܢܗ ܀ ⁷ ܡܪܝܐ ܐܡܪ ܠܝ ܕܒܪܝ ܐܢܬ : ܘܐܢܐ ܝܘܡܢܐ ܝܠܕܬܟ ܀ ⁸ ܫܐܠ
ܡܢܝ ܘܐܬܠ ܠܟ : ܥܡܡܐ ܠܝܪܬܘܬܟ : ܘܐܘܚܕܢܟ ܥܕܡܐ ܠܣܘܦܝܗ ܕܐܪܥܐ ܀ ⁹ ܬܪܥܐ
ܐܢܘܢ ܒܫܒܛܐ ܕܦܪܙܠܐ : ܘܐܝܟ ܡܐܢܐ ܕܦܚܪܐ ܬܫܪܐ ܐܢܘܢ ܀ ¹⁰ ܗܫܐ ܡܠܟܐ
ܐܣܬܟܠܘ : ܘܐܬܪܕܘ ܕܝܢܝܗ ܕܐܪܥܐ ܀ ¹¹ ܦܠܘܚܘ ܠܡܪܝܐ ܒܕܚܠܬܐ
ܘܐܫܬܘܕܥܘ ܒܪܬܠܐ ܀ ¹² ܢܫܩܘ ܒܪܐ ܕܠܐ ܢܪܓܙ : ܘܬܐܒܕܘܢ ܡܢ ܐܘܪܚܗ :
ܡܛܠ ܕܩܠܝܠ ܡܬܚܡܐ ܪܘܓܙܗ : ܛܘܒܝܗܘܢ ܠܟܠܗܘܢ ܕܬܟܝܠܝܢ ܥܠܘܗܝ ܀

Chapter I.

¹ ܘܗܘܐ ܦܬܓܡܗ ܕܡܪܝܐ ܥܠ ܝܘܢܢ ܒܪ ܐܡܬܝ ܠܡܐܡܪ. ² ܩܘܡ ܐܙܠ ܠܢܝܢܘܐ ܡܕܝܢܬܐ ܪܒܬܐ ܘܐܟܪܙ ܒܓܘܗ. ܡܛܠ ܕܣܠܩܬ ܒܝܫܘܬܗܘܢ ܩܕܡܝ. ³ ܘܩܡ ܝܘܢܢ ܠܡܥܪܩ ܠܬܪܫܝܫ ܡܢ ܩܕܡ ܡܪܝܐ. ܘܢܚܬ ܠܝܘܦܐ ܘܐܫܟܚ ܐܠܦܐ ܕܐܙܠܐ ܠܬܪܫܝܫ. ܘܝܗܒ ܐܓܪܐ ܘܢܚܬ ܠܗ ܠܡܐܙܠ ܥܡܗܘܢ ܠܬܪܫܝܫ ܡܢ ܩܕܡ ܡܪܝܐ. ⁴ ܘܡܪܝܐ ܐܪܡܝ ܪܘܚܐ ܪܒܬܐ ܒܝܡܐ ܘܗܘܐ ܡܥܠܘܠܐ ܪܒܐ ܒܝܡܐ. ܘܐܠܦܐ ܡܬܚܫܒܐ ܗܘܬ ܠܡܬܬܒܪܘ. ⁵ ܘܕܚܠܘ ܡܠܚܐ ܘܩܥܘ ܐܢܫ ܠܘܬ ܐܠܗܗ. ܘܫܕܘ ܡܐܢܐ ܡܢ ܐܠܦܐ ܕܒܗ ܠܝܡܐ ܕܢܩܠ ܡܢܗܘܢ. ܝܘܢܢ ܕܝܢ ܢܚܬ ܗܘܐ ܠܫܘܠܝܐ ܕܐܠܦܐ ܘܕܡܟ. ⁶ ܘܩܪܒ ܠܘܬܗ ܪܒ ܡܠܚܐ ܘܐܡܪ ܠܗ. ܡܢܐ ܕܡܟ ܐܢܬ ܩܘܡ ܩܪܝ ܠܐܠܗܟ ܕܠܡܐ ܢܦܨܝܢ ܐܠܗܐ ܘܠܐ ܢܐܒܕ. ⁷ ܘܐܡܪ ܓܒܪ ܠܚܒܪܗ. ܬܘ ܢܪܡܐ ܦܣܐ. ܘܢܕܥ ܡܛܠ ܡܢܐ ܗܘܐ ܗܢܐ ܒܝܫܬܐ ܥܠܝܢ. ܘܐܪܡܝܘ ܦܣܐ ܘܢܦܠ ܦܣܐ ܥܠ ܝܘܢܢ. ⁸ ܘܐܡܪܘ ܠܗ. ܡܛܠ ܡܢܐ ܒܝܫܬܐ ܗܘܐ ܡܬܚܫܒܢܘܬܐ ܐܬܬ ܥܠܝܢ. ܡܢܘ ܥܒܕܟ. ܘܡܢ ܐܝܟܐ ܐܬܝܬ. ܘܐܝܕܐ ܗܝ ܐܪܥܟ. ܘܡܢ ܐܝܢܐ ܥܡܐ ܐܢܬ. ⁹ ܐܡܪ ܠܗܘܢ ܥܒܪܝܐ ܐܢܐ. ܘܠܡܪܝܐ ܐܠܗܐ ܕܫܡܝܐ ܕܚܠ ܐܢܐ ܕܥܒܕ ܝܡܐ ܘܝܒܫܐ. ¹⁰ ܘܕܚܠܘ ܓܒܪܐ ܕܚܠܬܐ ܪܒܬܐ ܘܐܡܪܘ ܠܗ.

ܡܪܝܐ ܥܪܝܩ ܝܠܗ ܩܡܚܘܝܠܗ ܕܝܕܥܝ ܐ̈ܢܫܐ ܗ̈ܢܝ ܕܡܢ ܩܕܡ ܡܪܝܐ ܥܪܝܩ܀ 11 ܘܡܪܝ
ܠܗ ܐܝܟܢ ܐܡܥܒܕܝܚ ܠܘܟ ܕܝܡܐ ܢܬܢܝܚ ܡܢܢ ܣܒܒ ܕܝܡܐ ܛܠܝܐ ܝܠܗ
ܘܡܬܠܚܫܐ ܠܒܝܫ܀ 12 ܐܡܪ ܠܗܘܢ ܫܩܘܠܘ ܠܝ. ܘܫܕܘܠܝ
ܠܝܡܐ. ܘܝܡܐ ܢܬܢܝܚ ܡܢܘܟܘܢ ܣܒܒ ܕܝܕܥ ܝܢܐ ܕܒܥܠܬܝ ܥܠܥܠܐ ܪܒܐ ܡܢܬܠܝܗ
ܗܘܐ ܠܘܟܘܢ܀ 13 ܘܐܬܘܠܗ ܐ̈ܢܫܐ ܗܢܝ ܕܡܗܦܟܝ ܠܣܦܝܢܬܐ ܠܐ
ܝܒܫܐ. ܡܛܠ ܕܝܡܐ ܛܠܝܐ ܗܘܐ ܘܡܬܠܚܫܐ ܠܒܝܫܘܗܝ܀ 14 ܘܩܪܝܘ ܠܡܪܝܐ
ܘܐܡܪܘ. ܐܘ ܡܪܝܐ ܠܐ ܢܐܒܕ ܒܢܦܫܗ ܕܓܒܪܐ ܗܢܐ ܘܠܐ ܬܝܗܒ ܠܥܠܢ ܕܡܐ
ܙܟܝܐ ܡܛܠ ܕܐܢܬ ܗܘ ܡܪܝܐ ܕܐܝܟ ܕܨܒܝܬ ܥܒܕܬ܀ 15 ܘܫܩܠܘܗܝ
ܠܝܘܢܢ ܘܫܕܘܗܝ ܠܝܡܐ. ܘܩܡ ܝܡܐ ܡܢ ܙܥܦܗ܀
16 ܘܕܚܠܘ ܐ̈ܢܫܐ ܗ̈ܢܝ ܕܚܠܬܐ ܪܒܬܐ ܡܢ ܩܕܡ ܡܪܝܐ ܘܕܒܚܘ ܕܒܚܐ ܠܡܪܝܐ
ܘܢܕܪܘ ܢܕܪ̈ܐ܀

Chapter II.

1 ܘܡܢܝ ܡܪܝܐ ܢܘܢܐ ܪܒܐ ܕܒܠܥܗ ܠܝܘܢܢ. ܘܗܘܐ ܝܘܢܢ ܒܟܪܣܗ
ܕܢܘܢܐ ܬܠܬܐ ܐܝܡܡܝܢ ܘܬܠܬܐ ܠܠܝܢ܀ 2 ܘܨܠܝ ܝܘܢܢ ܩܕܡ ܡܪܝܐ
ܐܠܗܗ ܡܢ ܟܪܣܗ ܕܢܘܢܐ ܘܐܡܪ܀ 3 ܩܪܝܬ ܠܡܪܝܐ ܒܐܘܠܨܢܝ ܘܥܢܝܢܝ.
ܘܡܢ ܟܪܣܗ ܕܫܝܘܠ ܩܥܝܬ ܘܫܡܥܬ ܩܠܝ܀ 4 ܘܐܪܡܝܬܢܝ ܠܥܘܡܩܐ
ܕܠܒܗ ܕܝܡܐ ܘܢܗܪܐ ܚܕܪܢܝ. ܟܠܗܘܢ ܡܫܒܪ̈ܝܟ ܘܓܠܠ̈ܝܟ ܥܠܝ
ܥܒܪܘ܀ 5 ܐܢܐ ܕܝܢ ܐܡܪܬ ܕܐܬܛܪܕܬ ܡܢ ܩܕܡ ܥܝܢ̈ܝܟ. ܒܪܡ ܬܘܒ

ܐܢܐ ܠܥܘܡܩܐ ܕܠܒܐ ܕܝܡܐ ⁘ ⁴ ܢܗܪܘܬܐ ܡܢܝ ܚܕܪܝ ܠܝ. ܘܟܠܗܘܢ
ܡܘܓܠܝ. ܘܫܒܪܘܗ ܕܝܡܐ ܥܠܝ ܥܒܪ ⁘ ⁵ ܘܐܢܐ ܐܡܪܢ ܕܛܪܝܕܐ ܝܘܢ.
ܘܐܪܐ ܐܗܕܝ ܡܚܙܝܢܗ ܗܝܟܠ ܩܘܕܫܟ. ܘܐܦܩܘܢ ܡܝܐ ܡܢ ܚܕܪܐ ܢܦܫܐ
ܠܬܗܘܡܐ ⁘ ⁶ ܥܡ ܐܬܟܪܟܬ ܥܠܝ ܠܥܘܡܩܐ ܐܬܥܒܪܬ. ܘܓܠܠܐ ܡܚܡܢ ܪܝܫܝ
ܠܥܘܡܩܐ ܕܡܝܐ ⁘ ⁷ ܥܠ ܫܬܐܣܝ ܕܛܘܪܐ ܢܚܬܬ ܡܢ ܐܪܥܐ ܡܫܟܢܘܬܝ
ܥܠܡܝܢ ⁘ ¹⁰ ܐܢܐ ܕܝܢ ܒܩܠܐ ܕܬܘܕܝܬܐ ܐܕܒܚ ܠܟ. ܘܡܕܡ ܕܢܕܪܬ ܐܦܪܥ
ܦܘܪܩܢܐ ܠܡܪܝܐ ⁘ ¹¹ ܘܐܡܪ ܡܪܝܐ ܠܢܘܢܐ ܘܐܦܩܗ ܠܝܘܢܢ ܠܝܒܫܐ ⁘

Chapter III.

¹ ܘܗܘܐ ܦܬܓܡܗ ܕܡܪܝܐ ܥܠ ܝܘܢܢ ܕܬܪܬܝܢ ܙܒܢܝܢ ܠܡܐܡܪ ⁘
² ܩܘܡ ܐܙܠ ܠܢܝܢܘܐ ܡܕܝܢܬܐ ܪܒܬܐ ܘܐܟܪܙ ܒܗ ܟܪܘܙܘܬܐ ܕܐܡܪ ܐܢܐ ܠܟ ⁘
³ ܘܩܡ ܝܘܢܢ ܘܐܙܠ ܠܢܝܢܘܐ ܐܝܟ ܡܠܬܗ ܕܡܪܝܐ ܘܢܝܢܘܐ ܡܕܝܢܬܐ ܗܘܬ
ܪܒܬܐ ܠܐܠܗܐ ܡܪܕܝܐ ܕܬܠܬܐ ܝܘܡܝܢ ⁘ ⁴ ܘܫܪܝ ܝܘܢܢ ܠܡܥܠ ܠܡܕܝܢܬܐ
ܡܪܕܝܐ ܕܝܘܡܐ ܚܕ ܘܐܟܪܙ ܘܐܡܪ. ܥܘܕ ܐܪܒܥܝܢ ܝܘܡܝܢ ܢܝܢܘܐ ܡܬܗܦܟܐ ⁘
⁵ ܘܗܝܡܢܘ ܐܢܫܝܗ ܕܢܝܢܘܐ ܒܐܠܗܐ ܘܩܪܘ ܨܘܡܐ ܘܠܒܫܘ ܣܩܐ ܡܢ
ܪܘܪܒܢܝܗܘܢ ܘܥܕܡܐ ܠܙܥܘܪܝܗܘܢ ⁘ ⁶ ܘܡܛܬ ܡܠܬܐ ܠܡܠܟܐ ܕܢܝܢܘܐ ܘܩܡ
ܡܢ ܟܘܪܣܝܗ ܘܫܠܚ ܠܒܘܫܗ ܡܢܗ. ܘܟܣܝ ܣܩܐ ܘܝܬܒ ܥܠ ܩܛܡܐ ⁘
⁷ ܘܐܟܪܙܘ ܘܐܡܪܘ ܒܢܝܢܘܐ . ܡܢ ܦܘܩܕܢܐ ܕܡܠܟܐ ܘܕܪܘܪܒܢܘܗܝ ܠܡܐܡܪ

ܘܒܥܝܪܐ ܘܬܘܪܐ ܘܥܢܐ ܠܐ ܢܛܥܡܘܢ ܡܕܡ ܘܠܐ ܢܪܥܘܢ ܐܦܠܐ ܡܝܐ ܢܫܬܘܢ ܀ 8 ܐܠܐ
ܢܬܟܣܘܢ ܣܩܐ ܒܢܝܢܫܐ ܘܒܥܝܪܐ. ܘܢܩܪܘܢ ܠܐܠܗܐ ܒܚܝܠܐ ܘܢܬܦܢܘܢ ܐܢܫ
ܡܢ ܐܘܪܚܗ ܒܝܫܬܐ ܘܡܢ ܚܛܘܦܝܐ ܕܐܝܬ ܒܐܝܕܝܗܘܢ ܀ 9 ܡܢܘ ܝܕܥ ܐܢ
ܡܬܦܢܐ ܐܠܗܐ ܘܡܬܪܚܡ ܥܠܝܢ ܘܡܗܦܟ ܡܢ ܚܡܬܐ ܕܪܘܓܙܗ ܘܠܐ ܢܐܒܕ ܀
10 ܘܚܙܐ ܐܠܗܐ ܥܒܕܝܗܘܢ ܕܗܦܟܘ ܡܢ ܐܘܪܚܬܗܘܢ ܒܝܫܬܐ ܘܐܗܦܟ ܡܢܗܘܢ
ܒܝܫܬܐ ܕܪܘܓܙܗ ܘܠܐ ܐܝܬܝ ܐܢܝܢ ܀

Chapter IV.

1 ܘܒܐܫ ܠܝܘܢܢ ܒܝܫܬܐ ܪܒܬܐ ܘܐܬܟܪܝ ܠܗ ܛܒ ܀ 2 ܘܨܠܝ ܩܕܡ
ܡܪܝܐ ܘܐܡܪ ܐܘ ܡܪܝܐ ܠܐ ܗܘܬ ܗܕܐ ܡܠܬܝ ܟܕ ܐܝܬܝ ܒܐܪܥܝ ܘܡܛܠ ܗܢܐ
ܩܕܡܬ ܗܘܝܬ ܠܡܥܪܩ ܠܬܪܫܝܫ ܝܕܥ ܗܘܝܬ ܓܝܪ ܕܐܠܗܐ ܐܢܬ ܡܚܢܢܐ
ܘܡܪܚܡܢܐ ܘܢܓܝܪ ܪܘܚܐ ܘܣܓܝ ܛܝܒܘܬܐ ܘܡܬܬܘܐ ܐܢܬ ܥܠ ܒܝܫܬܐ ܀
3 ܘܗܫܐ ܡܪܝܐ ܣܒ ܢܦܫܝ ܡܢܝ ܡܛܠ ܕܛܒ ܠܝ ܕܐܡܘܬ ܡܢ
ܕܐܚܐ ܀ 4 ܘܐܡܪ ܠܗ ܡܪܝܐ. ܛܒ ܐܬܟܪܝ ܠܟ ܀ 5 ܘܢܦܩ ܝܘܢܢ ܠܒܪ
ܡܢ ܡܕܝܢܬܐ ܘܝܬܒ ܠܗ ܡܢ ܡܕܢܚܗ ܕܡܕܝܢܬܐ ܘܥܒܕ ܠܗ ܬܡܢ ܡܛܠܬܐ
ܘܝܬܒ ܬܚܘܬܗ ܒܛܠܠܐ ܕܢܚܙܐ ܡܢܐ ܗܘܐ ܠܗ ܠܡܕܝܢܬܐ ܀ 6 ܘܡܢܝ
ܡܪܝܐ ܐܠܗܐ ܠܩܪܐ ܘܣܠܩܬ ܠܥܠ ܡܢ ܝܘܢܢ. ܘܗܘܬ ܛܠܠܐ
ܥܠ ܪܫܗ. ܘܐܦܨܝ ܠܗ ܡܢ ܒܝܫܬܗ. ܘܚܕܝ ܝܘܢܢ ܒܩܪܐ ܕܚܕܘܬܐ

ܚܕܘܬܐ ܪܒܬܐ ܀ ⁷ ܘܬܘܠܥܬܐ ܐܙܕܡܢܬ ܥܡ ܡܕܢܚܐ ܕܨܦܪܐ ܠܡܚܪܬܗ ܘܡܚܬܗ

ܠܩܪܐ . ܘܡܚܬܗ ܠܥܩܪܐ ܕܩܪܐ ܘܝܒܫܬܗ ܀ ⁸ ܘܟܕ ܕܢܚ ܫܡܫܐ ܦܩܕ

ܡܪܝܐ ܐܠܗܐ ܠܪܘܚܐ ܕܩܕܝܡܐ ܘܐܫܚܢܬܗ ܠܐܪܥܐ . ܘܡܚܬ ܫܡܫܐ ܥܠ ܪܝܫܗ

ܕܝܘܢܢ . ܘܐܬܢܓܕ ܘܫܐܠ ܡܘܬܐ ܠܢܦܫܗ ܘܐܡܪ . ܛܒ ܠܝ ܕܐܡܘܬ ܡܢ

ܠܡܫܬܡܠܘ ܢܦܫܝ ܡܢ ܚܝܐ ܕܠܐ ܐܝܬ ܛܒ ܐܢܐ ܡܢ ܐܚܝܐ ܀ ⁹ ܘܐܡܪ

ܡܪܝܐ ܐܠܗܐ ܠܝܘܢܢ . ܫܦܝܪ ܚܡܬ ܠܟ ܥܠ ܥܩܪܐ ܕܩܪܐ . ܘܐܡܪ ܝܘܢܢ .

ܛܒ ܚܡܬ ܠܝ ܥܕܡܐ ܠܡܘܬܐ ܀ ¹⁰ ܐܡܪ ܠܗ ܡܪܝܐ . ܐܢܬ ܚܣܬ ܥܠ

ܩܪܐ ܕܠܐ ܠܐܝܬ ܒܗ ܘܠܐ ܪܒܝܬܝܗܝ ܕܒܪ ܠܠܝܗ ܗܘܐ ܘܒܪ ܠܠܝܗ

ܐܒܕ ܀ ¹¹ ܐܢܐ ܕܝܢ ܠܐ ܐܚܘܣ ܥܠ ܢܝܢܘܐ ܡܕܝܢܬܐ ܪܒܬܐ . ܕܐܝܬ ܒܗ ܣܓܝ

ܡܢ ܬܪܬܥܣܪܐ ܪܒܘܢ ܒܢܝܢܫܐ ܕܠܐ ܝܕܥܝܢ ܒܝܬ ܝܡܝܢܗܘܢ ܠܣܡܠܗܘܢ

ܘܒܥܝܪܐ ܣܓܝܐܐ ܀

THE PROPHECY OF MALACHI.

Chapter I.

¹ ܡܫܩܠܐ ܕܦܬܓܡܘܗܝ ܕܡܪܝܐ ܥܠ ܐܝܣܪܝܠ ܒܝܕ ܡܠܐܟܝ ܀ ² ܪܚܡܬܟܘܢ
ܐܡܪ ܡܪܝܐ ܘܐܡܪܝܬܘܢ ܐܢܬܘܢ ܒܡܢܐ ܪܚܡܬܢ. ܠܡܢܐ ܥܣܘ ܠܐ ܗܘܐ ܐܚܘܗܝ
ܕܝܥܩܘܒ ܐܡܪ ܡܪܝܐ. ܘܪܚܡܬܗ ܠܝܥܩܘܒ ܘܠܥܣܘ ܣܢܝܬܗ ܀ ³ ܘܥܒܕܬ
ܛܘܪ̈ܘܗܝ ܠܚܘܪܒܐ ܘܝܪܬܘܬܗ ܠܬܢܝܢ̈ܐ ܕܡܕܒܪܐ ܀ ⁴ ܘܐܢ ܢܐܡܪܘܢ ܐܕܘܡܝ̈ܐ.
ܐܬܡܣܟܢܢ ܢܬܘܒ ܢܒܢܐ ܚܪ̈ܒܬܐ. ܗܟܢܐ ܐܡܪ ܡܪܝܐ. ܗܢܘܢ ܢܒܢܘܢ ܘܐܢܐ
ܐܣܚܘܦ. ܘܢܩܪܘܢ ܐܢܘܢ ܬܚܘܡܐ ܕܪܫܝܥܘܬܐ. ܘܥܡܐ ܕܪܓܙ ܥܠܘܗܝ ܡܪܝܐ
ܠܥܠܡ ܀ ⁵ ܘܥܝܢ̈ܝܟܘܢ ܢܚܙ̈ܝܢ. ܘܐܢܬܘܢ ܬܐܡܪܘܢ. ܢܬܪܘܪܒ ܡܪܝܐ ܠܥܠ
ܡܢ ܬܚܘܡܐ ܕܐܝܣܪܝܠ ܀ ⁶ ܒܪܐ ܡܝܩܪ ܠܐܒܘܗܝ ܘܥܒܕܐ ܠܡܪܗ. ܐܢ ܐܒܐ
ܐܢܐ ܐܝܟܘ ܕܡܝܩܪܝܢ ܐܢܬܘܢ ܠܝ. ܘܐܢ ܡܪܐ ܐܢܐ ܐܝܟܘ ܕܕܚܠܝܢ ܐܢܬܘܢ ܡܢܝ.
ܐܡܪ ܡܪܝܐ ܚܝܠܬܢܐ ܠܟܘܢ ܟܘܗܢ̈ܐ ܕܫܛܝܢ ܐܢܬܘܢ ܫܡܝ. ܘܐܢ ܐܡܪܝܢ ܐܢܬܘܢ.
ܒܡܢܐ ܐܫܛܢ ܫܡܟ ܀ ⁷ ܥܠ ܕܡܩܪܒܝܢ ܐܢܬܘܢ ܥܠ ܡܕܒܚܝ ܠܚܡܐ ܛܢܦܐ.
ܘܐܢ ܐܡܪܝܢ ܐܢܬܘܢ ܒܡܢܐ ܛܢܦܢܟ. ܒܐܡܪܝܢ ܐܢܬܘܢ ܕܫܝܛ ܗܘ ܦܬܘܪܗ
ܕܡܪܝܐ ܀ ⁸ ܘܕܡܩܪܒܝܢ ܐܢܬܘܢ ܥܘܝܪܐ ܠܡܕܒܚܐ ܕܒܚܬܐ ܠܐ ܒܝܫ.
ܘܡܩܪܒܝܢ ܐܢܬܘܢ ܚܓܝܪܐ ܘܟܪܝܗܐ ܕܒܚܬܐ ܠܐ ܒܝܫ. ܩܪܒܝܗܝ ܗܐ
ܠܫܠܝܛܟ ܕܒܚܬܐ ܐܢ ܢܪܥܐ ܒܟ ܐܘ ܢܣܒ ܐܦܝܟ ܐܡܪ ܡܪܝܐ ܚܝܠܬܢܐ ܀

[9] ܡܟܝܠ ܒܥܘ ܦܨܘܗ ܕܐܠܗܐ. ܘܚܢܢ ܥܠܢ. ܕܡܢ ܐܝܕܘܟܘܢ ܗܘܐ ܗܕܐ. ܠܐ
ܐܝܟ ܦܐܬܘܟܘܢ ܐܡܪ ܡܪܝܐ ܚܝܠܬܢܐ܀ [10] ܡܢܘ ܐܝܬ ܒܟܘܢ ܕܩܐܪܘ ܬܪ̈ܥܐ.
ܠܐ ܡܫܘܓܪܝܬܘܢ ܡܓܢ ܢܘܪܐ ܕܡܕܒܚܝ. ܠܐ ܨܒܝܐ ܐܝܠܝ ܒܟܘܢ ܐܡܪ ܡܪܝܐ ܚܝܠܬܢܐ.
ܘܩܘܪܒܢܐ ܠܐ ܩܒܠ ܡܐܝܕܟܘܢ܀ [11] ܡܛܠ ܕܡܢ ܡܕܢܚܗ ܕܫܡܫܐ ܘܗܡܐ
ܠܡܥܪܒܗ ܪܒܐ ܝܠܗ ܫܡܝ ܒܥܡ̈ܡܐ. ܘܒܟܠ ܐܬܪ ܩܛܪܝܢ ܒܣ̈ܡܐ.
ܘܡܩܪܒܝܢ ܠܫܡܝ ܩܘܪܒܢܐ ܕܟܝܐ. ܡܛܠ ܕܪܒܐ ܝܠܗ ܫܡܝ ܒܥܡ̈ܡܐ ܐܡܪ
ܡܪܝܐ ܚܝܠܬܢܐ܀ [12] ܘܐܢܬܘܢ ܡܛܢܦܝܬܘܢ ܐܠܗ ܒܐܡܪܝܢ ܐܢܬܘܢ.
ܕܦܬܘܪܗ ܕܡܪܝܐ ܡܛܢܦ ܝܠܗ. ܘܐܝܡܢ ܐܟܠ ܡܐܟܘܠܬܗ܀ [13] ܘܐܡܪܝܢ
ܐܢܬܘܢ. ܗܐ ܡܢ ܥܡܠܐ ܝܠܗ. ܘܢܦܚܝܬܘܢ ܒܗ ܐܡܪ ܡܪܝܐ ܚܝܠܬܢܐ. ܘܡܝܬܝܬܘܢ
ܐܢܬܘܢ ܡܢ ܓܙܠܬܐ ܩܘܪܒܢܐ ܚܓܝܪܐ ܘܟܪܝܗܐ. ܠܐ ܡܩܒܠܢܐ ܠܗܘܢ ܡܢ
ܐܝܕܘܟܘܢ ܐܡܪ ܡܪܝܐ ܚܝܠܬܢܐ܀ [14] ܘܠܝܛܐ ܗܘ ܕܐܝܬ ܠܗ ܒܡܪܥܝܬܗ ܕܟܪܐ.
ܘܢܕܪ ܘܡܕܒܚ ܕܒܝܚܐ ܠܡܪܝܐ. ܡܛܠ ܕܡܠܟܐ ܐܢܐ ܪܒܐ ܐܡܪ ܡܪܝܐ
ܚܝܠܬܢܐ ܘܫܡܝ ܕܚܝܠܐ ܒܥܡ̈ܡܐ܀

Chapter II.[4]

[1] ܘܗܫܐ ܦܘܩܕܢܐ ܗܢܐ ܠܟܘܢ ܝܠܗ ܟܗ̈ܢܐ܀ [2] ܐܢ ܠܐ ܫܡܥܝܬܘܢ ܘܐܢ ܠܐ
ܣܝܡܝܬܘܢ ܒܠܒܘܟܘܢ ܠܡܝܗܒ ܐܝܩܪܐ ܠܫܡܝ. ܐܡܪ ܡܪܝܐ ܚܝܠܬܢܐ. ܐܫܕܪ
ܥܠܘܟܘܢ ܠܘܛܬܐ ܘܠܝܛ ܒܘܪܟܬܟܘܢ ܡܛܠ ܕܠܐ ܣܝܡܝܬܘܢ ܠܗ ܒܠܒܘܟܘܢ܀

³ ܗܐ ܓܥܪ ܐܢܐ ܒܙܪܥܐ ܕܪܥܐ. ܘܐܕܪܐ ܦܪܬܐ ܥܠ ܐܦ̈ܝܟܘܢ. ܘܦܪܬܐ ܥܠ
ܥܐܕ̈ܝܟܘܢ. ܘܐܫܩܘܠܟܘܢ ܥܡܗ ⁖ ⁴ ܘܬܕܥܘܢ ܕܫܕܪܬ ܠܟܘܢ ܦܘܩܕܢܐ ܗܢܐ.
ܕܬܗܘܐ ܕܝܬܝܩܝ ܥܡ ܠܘܝ ܐܡܪ ܡܪܝܐ ܚܝܠܬܢܐ ⁖ ⁵ ܩܝܡܝ ܗܘܐ ܥܡܗ.
ܚ̈ܝܐ ܘܫܠܡܐ ܝܗܒܬ ܐܢܘܢ ܠܗ ܕܢܕܚܠ· ܘܕܚܠ ܡܢܝ ܘܡܢ ܩܕܡ ܫܡܝ ܙܥ
ܗܘܐ ⁖ ⁶ ܢܡܘܣܐ ܕܩܘܫܬܐ ܗܘܐ ܒܦܘܡܗ. ܘܥܘܠܐ ܠܐ ܐܫܬܟܚ ܒܣܦܘ̈ܬܗ.
ܒܫܠܡܐ ܘܒܬܪܝܨܘܬܐ ܗܠܟ ܥܡܝ. ܘܣܓܝ̈ܐܐ ܐܗܦܟ ܡܢ ܥܘܠܐ ⁖ ⁷ ܡܛܠ
ܕܣܦܘ̈ܬܗ ܕܟܗܢܐ ܢܛܪܢ ܝܕܥܬܐ. ܘܢܡܘܣܐ ܢܒܥܘܢ ܡܢ ܦܘܡܗ. ܡܛܠ
ܕܡܠܐܟܗ ܗܘ ܕܡܪܝܐ ܚܝܠܬܢܐ ⁖ ⁸ ܐܢܬܘܢ ܕܝܢ ܣܛܝܬܘܢ ܡܢ ܐܘܪܚܐ.
ܘܐܟܫܠܬܘܢ ܠܣܓܝ̈ܐܐ ܡܢ ܢܡܘܣܐ. ܘܚܒܠܬܘܢ ܩܝܡܐ ܕܠܘܝ ܐܡܪ ܡܪܝܐ
ܚܝܠܬܢܐ ⁖ ⁹ ܐܦ ܐܢܐ ܝܗܒܬܟܘܢ ܫܝܛܝܢ ܘܡܟܝܟܝܢ ܠܟܠ ܥܡܐ. ܕܠܐ
ܢܛܪܬܘܢ ܐܘܪܚܬܝ. ܘܢܣܒܬܘܢ ܒܐܦܐ ܒܢܡܘܣܐ ⁖ ¹⁰ ܘܠܡܢܐ ܠܐ ܗܘܐ ܚܕ ܐܒܐ
ܠܟܠܢ. ܐܘ ܠܐ ܗܘܐ ܚܕ ܐܠܗܐ ܒܪܢ. ܠܡܢܐ ܡܫܥܒܕܝܢܢ ܓܒܪ ܒܐܚܘܗܝ
ܘܡܣܠܝܢܢ ܩܝܡܐ ܕܐܒܗ̈ܬܢ ⁖ ¹¹ ܕܓܠ ܝܗܘܕܐ ܘܛܢܦܘܬܐ ܐܬܥܒܕܬ
ܒܐܝܣܪܝܠ ܘܒܐܘܪܫܠܡ. ܡܛܠ ܕܚܠܠ ܝܗܘܕܐ ܩܘܕܫܗ ܕܡܪܝܐ ܕܪܚܡ.
ܘܪܚܡ ܘܢܣܒ ܠܒܢܬܐ ܢܘܟܪܝܬܐ ⁖ ¹² ܢܘܒܕ ܡܪܝܐ ܠܓܒܪܐ ܕܥܒܕ ܗܕܐ.
ܘܠܒܪܗ ܘܠܒܪ ܒܪܗ ܡܢ ܡܫܟܢܘܗܝ ܕܝܥܩܘܒ. ܘܠܐ ܢܗܘܐ ܠܗ ܕܡܩܪܒ
ܩܘܪܒܢܐ ܠܡܪܝܐ ܚܝܠܬܢܐ ⁖ ¹³ ܘܗܕܐ ܬܢܝܢܘܬ ܥܒܕܬܘܢ. ܟܣܝܬܘܢ ܕܡ̈ܥܐ
ܠܡܕܒܚܗ ܕܡܪܝܐ. ܒܒܟܝܐ ܘܬܢܚܬܐ. ܡܛܠ ܕܠܐ ܡܬܦܢܐ ܥܠ ܩܘܪܒܢܝܟܘܢ ܘܠܐ
ܡܩܒܠ ܠܗ ܠܪܥܘܬܐ ܡܢ ܐܝܕ̈ܝܟܘܢ ⁖ ¹⁴ ܘܐܢ ܐܡܪܝܢ ܐܢܬܘܢ. ܥܠ ܡܢܐ.

ܟܠ ܕܐܥܒܕ ܡܪܝܐ ܒܥܠܡܝ ܠܐܝܬܐ ܡܫܟܚܬܝ. ܘܐܝܬ ܕܒܓܕܫܐ ܒܗ ܗܘܝ
ܡܫܬܘܬܝ ܘܐܝܬܐ ܕܩܝܡܬܝ. 15 ܕܠܡܐ ܠܐ ܗܘܐ ܚܕ ܥܒܕ ܘܫܪܟܐ ܕܪܘܚܐ
ܕܝܠܗ ܗܘ. ܚܕ ܒܢܐ ܐܪܐ ܡܢ ܐܠܗܐ. ܐܢܛܪܘ ܒܪܘܚܟܘܢ ܘܐܢܫ ܒܐܝܬܐ
ܡܫܟܚܬܗ ܠܐ ܢܒܓܠ. 16 ܐܡܪ ܡܪܝܐ ܡܫܒܚܬܐ ܐܠܗܐ ܕܐܝܣܪܐܝܠ. ܘܟܘܠܐ
ܒܡܪܝܚܘܬܗ ܠܐ ܢܚܒܐ ܐܡܪ ܡܪܝܐ ܡܫܒܚܬܐ. ܐܢܛܪܘ ܒܪܘܚܟܘܢ ܘܠܐ ܬܒܓܠܘܢ.
17 ܐܠܐܝܬܘܢ ܠܡܪܝܐ ܒܡܠܘܟܘܢ. ܘܐܢ ܐܡܪܝܬܘܢ ܐܝܟܐ ܐܠܐܝܬܝܢܢ.
ܒܐܡܪܝܬܘܢ ܕܟܠ ܕܥܒܕ ܒܝܫ ܒܥܝܢܐ ܡܪܝܐ. ܘܒܗܘܢ ܗܘ
ܨܒܐ ܘܐܝܟܐ ܐܠܗܐ ܕܕܝܢܐ.

Chapter III.

1 ܗܐ ܐܢܐ ܫܕܪ ܐܢܐ ܡܠܐܟܝ. ܘܒܢܐ ܐܘܪܚܐ ܩܕܡܝ. ܘܡܢ ܫܠܝܐ ܒܝܬܐ
ܠܗܝܟܠܗ ܕܡܪܝܐ ܕܒܥܝܬܘܢ ܐܝܬܘܢ ܠܗ. ܘܡܠܐܟܐ ܕܩܝܡܐ ܕܨܒܝܬܘܢ ܐܝܬܘܢ
ܒܗ. ܗܐ ܐܬܐ ܐܡܪ ܡܪܝܐ ܡܫܒܚܬܐ. 2 ܡܢܘ ܡܣܝܒܪ ܝܘܡܐ ܕܐܬܐ ܒܗ ܐܘ
ܡܢܘ ܩܝܡ ܠܡܬܚܙܝ ܐܥܠܗ ܘܒܓܠܝܬܗ. ܣܒܒ ܕܐܝܟ ܢܘܪܐ ܗܘ ܕܨܪܦܐ ܘܐܝܟ
ܒܪܝܬܐ ܕܡܫܘܪܐ. 3 ܕܝܬܒ ܠܡܨܪܦ ܘܠܡܕܟܝܘ ܐܝܟ ܣܐܡܐ. ܘܢܕܟܐ
ܠܒܢܝ ܕܠܘܝ ܘܢܨܠܐ ܐܢܘܢ ܐܝܟ ܕܗܒܐ ܘܐܝܟ ܣܐܡܐ. ܘܢܗܘܘܢ ܡܩܪܒܝܢ
ܩܘܪܒܢܐ ܠܡܪܝܐ ܒܙܕܝܩܘܬܐ. 4 ܘܢܒܣܡ ܠܡܪܝܐ ܩܘܪܒܢܐ ܕܝܗܘܕܐ
ܘܐܘܪܫܠܡ ܐܝܟ ܝܘܡܬܐ ܕܡܢ ܥܠܡ. ܘܐܝܟ ܫܢܝܐ ܕܡܢ ܩܕܝܡ.

⁵ ܘܐܬܩܪܒ ܠܘܬܟܘܢ ܒܕܝܢܐ. ܘܐܗܘܐ ܣܗܕܐ ܡܣܪܗܒܐ ܒܚܪܫܐ ܘܒܓܝܪܐ ܘܒܐܝܠܝܢ ܕܝܡܝܢ ܒܫܡܝ ܠܕܓܠܘܬܐ. ܘܒܐܝܠܝܢ ܕܛܠܡܝܢ ܐܓܪܐ ܕܐܓܝܪܐ. ܘܡܣܟܢܐ ܘܝܬܡܐ ܘܐܪܡܠܬܐ. ܘܡܨܠܝܢ ܕܝܢܗ ܕܐܢܫ ܥܠ ܐܢܫܐ ܡܣܟܢܐ ܒܕܝܢ. ܘܠܐ ܕܚܠܘ ܡܢܝ. ܐܡܪ ܡܪܝܐ ܚܝܠܬܢܐ. ⁶ ܡܛܠ ܕܐܢܐ ܐܢܐ ܡܪܝܐ ܘܠܐ ܐܫܬܢܝܬ. ܘܐܢܬܘܢ ܒܢܝ ܝܥܩܘܒ ܠܐ ܚܒܠܬܘܢ ܡܢ ܩܕܡܝܗܘܢ. ⁷ ܡܢ ܝܘܡܝ ܐܒܗܝܟܘܢ ܣܛܝܬܘܢ ܡܢ ܦܘܩܕܢܝ. ܘܠܐ ܢܛܪܬܘܢ ܐܢܘܢ. ܐܬܦܢܘ ܠܘܬܝ ܘܐܬܦܢܐ ܠܘܬܟܘܢ ܐܡܪ ܡܪܝܐ ܚܝܠܬܢܐ. ⁸ ܘܐܢ ܐܡܪܝܢ ܐܢܬܘܢ ܒܡܢܐ ܢܬܦܢܐ. ܕܡܢܐ ܢܛܠܡ ܒܪܢܫܐ ܠܐܠܗܐ. ܐܝܟ ܕܐܢܬܘܢ ܛܠܡܝܢ ܐܢܬܘܢ ܠܝ. ܘܐܢ ܐܡܪܝܢ ܐܢܬܘܢ ܒܡܢܐ ܛܠܡܢܟ. ܒܡܥܣܪܐ ܘܒܦܪܝܫܘܬܐ. ⁹ ܒܠܘܛܬܐ ܡܬܠܛܝܢ ܐܢܬܘܢ ܘܠܝ ܛܠܡܝܢ ܐܢܬܘܢ. ¹⁰ ܐܥܠܘ ܟܠܗ ܡܥܣܪܐ ܠܒܝܬ ܐܘܨܪܐ. ܘܢܗܘܐ ܡܬܪܣܝܢܘܬܐ ܒܒܝܬܝ. ܘܢܣܘܢܝ ܒܗܕܐ ܐܡܪ ܡܪܝܐ ܚܝܠܬܢܐ ܕܐܦܬܚ ܠܟܘܢ ܟܘܐ ܕܫܡܝܐ ܘܐܫܘܕ ܠܟܘܢ ܒܘܪܟܬܐ. ܥܕܡܐ ܕܬܐܡܪܘܢ. ܣܓܝ. ¹¹ ܘܐܓܥܘܪ ܒܐܟܘܠܐ ܘܠܐ ܢܚܒܠ ܦܐܪܐ ܕܐܪܥܐ. ܘܠܐ ܬܬܚܪܒ ܠܟܘܢ ܓܦܬܐ ܕܒܚܩܠܐ ܐܡܪ ܡܪܝܐ ܚܝܠܬܢܐ. ¹² ܘܢܛܘܒܘܢܟܘܢ ܟܠܗܘܢ ܥܡܡܐ ܟܕ ܬܗܘܘܢ ܐܪܥܐ ܕܨܒܝܢܐ ܐܡܪ ܡܪܝܐ ܚܝܠܬܢܐ. ¹³ ܐܥܫܢܘ ܥܠܝ ܡܠܝܟܘܢ ܐܡܪ ܡܪܝܐ. ܘܐܢ ܐܡܪܝܢ ܐܢܬܘܢ. ܡܢܐ ܐܡܠܠܢ ܥܠܝܟ. ¹⁴ ܐܡܪܬܘܢ ܕܣܪܝܩܐܝܬ ܦܠܚܝܢܢ ܠܐܠܗܐ. ܘܡܢܐ ܐܘܬܪܢ ܕܢܛܪܢ ܢܛܘܪܬܗ. ܘܕܗܠܟܢ ܡܬܟܦܢܐܝܬ ܩܕܡ ܡܪܝܐ ܚܝܠܬܢܐ. ¹⁵ ܘܡܢ ܗܫܐ ܛܘܒܐ ܐܡܪܝܢ ܚܢܢ ܠܥܘܠܐ. ܘܡܬܒܢܝܢ ܥܒܕܝ ܪܘܫܥܐ. ܘܡܢܣܝܢ ܠܐܠܗܐ ܘܡܬܦܨܝܢ.

16 ܗܝܕܝܢ ܡܠܠܘ ܕܚܠܘܗܝ ܕܡܪܝܐ ܓܒܪ ܥܡ ܚܒܪܗ. ܘܨܬ ܡܪܝܐ ܘܫܡܥ.
ܘܟܬܒ ܒܣܦܪܐ ܕܕܘܟܪܢܐ ܩܕܡܘܗܝ ܠܕܚܠܘܗܝ ܘܠܐܝܠܝܢ ܕܡܚܫܒܝܢ
ܠܫܡܗ܀ 17 ܘܢܗܘܘܢ ܠܝ ܐܡܪ ܡܪܝܐ ܚܝܠܬܢܐ. ܠܝܘܡܐ ܕܥܒܕ ܐܢܐ
ܩܢܝܢܐ. ܘܐܚܘܣ ܥܠܝܗܘܢ ܐܝܟ ܕܚܐܣ ܓܒܪܐ ܥܠ ܒܪܗ ܕܦܠܚ ܠܗ܀
18 ܘܬܬܘܒܘܢ ܘܬܚܙܘܢ ܒܝܬ ܙܕܝܩܐ ܠܥܘܠܐ. ܒܝܬ ܕܦܠܚ ܠܐܠܗܐ ܠܐܝܠܝܢ
ܕܠܐ ܦܠܚܘܗܝ܀

Chapter IV.

1 ܡܛܠ ܕܗܐ ܝܘܡܐ ܐܬܐ ܘܝܩܕ ܐܝܟ ܬܢܘܪܐ. ܘܢܗܘܘܢ ܟܠܗܘܢ
ܥܘܠܐ ܘܟܠܗܘܢ ܥܒܕܝ ܥܘܠܐ ܩܫܐ. ܘܢܘܩܕ ܐܢܘܢ ܝܘܡܐ ܕܐܬܐ ܐܡܪ ܡܪܝܐ
ܚܝܠܬܢܐ. ܘܠܐ ܢܫܒܘܩ ܠܗܘܢ ܥܩܪܐ ܘܣܘܟܐ܀ 2 ܘܢܕܢܚ ܠܟܘܢ ܠܕܚܠܝ
ܫܡܝ ܫܡܫܐ ܕܙܕܝܩܘܬܐ. ܘܐܣܝܘܬܐ ܥܠ ܟܢܦܘܗܝ. ܘܬܦܩܘܢ ܘܬܕܘܨܘܢ ܐܝܟ
ܥܓܠܐ ܕܒܩܪܐ܀ 3 ܘܬܕܘܫܘܢ ܠܪܫܝܥܐ. ܡܛܠ ܕܢܗܘܘܢ ܩܛܡܐ ܬܚܝܬ ܦܪܣܬ
ܪܓܠܝܟܘܢ ܒܝܘܡܐ ܕܥܒܕ ܐܢܐ ܐܡܪ ܡܪܝܐ ܚܝܠܬܢܐ܀ 4 ܐܬܕܟܪܘ ܢܡܘܣܗ
ܕܡܘܫܐ ܥܒܕܝ ܕܦܩܕܬܗ ܒܚܘܪܒ. ܥܠ ܟܠܗ ܐܝܣܪܝܠ ܦܘܩܕܢܐ ܘܕܝܢܐ܀
5 ܗܐ ܡܫܕܪ ܐܢܐ ܠܟܘܢ ܠܐܝܠܝܐ ܢܒܝܐ ܥܕܠܐ ܕܢܐܬܐ ܝܘܡܗ ܕܡܪܝܐ ܪܒܐ
ܘܕܚܝܠܐ܀ 6 ܕܢܦܢܐ ܠܒܐ ܕܐܒܗܐ ܥܠ ܒܢܝܐ. ܘܠܒܐ ܕܒܢܝܐ ܥܠ ܐܒܗܝܗܘܢ.
ܕܠܐ ܐܬܐ ܘܐܡܚܝܘܗ ܠܐܪܥܐ ܠܓܡܪ܀

FROM THE GOSPEL OF ST. MATTHEW.

Chapter XXVI.

¹ ܘܗܘܐ ܕܟܕ ܫܠܡ ܝܫܘܥ ܟܠܗܘܢ ܡܠܐ ܗܠܝܢ ܐܡܪ ܠܬܠܡܝܕܘܗܝ ܀
² ܝܕܥܝܢ ܐܢܬܘܢ ܕܒܬܪ ܬܪܝܢ ܝܘܡܝܢ ܗܘܐ ܦܨܚܐ ܘܒܪܗ ܕܐܢܫܐ ܡܫܬܠܡ
ܕܢܙܕܩܦ ܀ ³ ܗܝܕܝܢ ܐܬܟܢܫܘ ܪܒܝ ܟܗܢܐ ܘܣܦܪܐ ܘܩܫܝܫܐ ܕܥܡܐ ܠܕܪܬܗ
ܕܪܒ ܟܗܢܐ ܕܡܬܩܪܐ ܩܝܦܐ ܀ ⁴ ܘܐܬܡܠܟܘ ܥܠ ܝܫܘܥ ܕܒܢܟܠܐ ܢܐܚܕܘܢܝܗܝ
ܘܢܩܛܠܘܢܝܗܝ ܀ ⁵ ܘܐܡܪܝܢ ܗܘܘ ܠܐ ܒܥܐܕܐ ܕܠܐ ܢܗܘܐ ܫܓܘܫܝܐ ܒܥܡܐ ܀
⁶ ܘܟܕ ܗܘܐ ܝܫܘܥ ܒܒܝܬ ܥܢܝܐ ܒܒܝܬܗ ܕܫܡܥܘܢ ܓܪܒܐ ܀ ⁷ ܩܪܒܬ ܠܗ
ܐܢܬܬܐ ܕܐܝܬ ܠܗ ܫܛܝܦܐ ܕܒܣܡܐ ܕܡܫܚܐ ܛܒܐ ܕܝܩܝܪ ܘܐܫܦܟܬܗ ܥܠ
ܪܝܫܗ ܕܝܫܘܥ ܟܕ ܣܡܝܟ ܀ ⁸ ܚܙܘ ܕܝܢ ܬܠܡܝܕܘܗܝ ܘܐܬܒܐܫ ܠܗܘܢ
ܘܐܡܪܝܢ ܠܡܢܐ ܐܒܕܢܐ ܗܢܐ ܀ ⁹ ܡܫܟܚ ܗܘܐ ܓܝܪ ܕܢܙܕܒܢ ܗܢܐ ܒܣܓܝ
ܘܢܬܝܗܒ ܠܡܣܟܢܐ ܀ ¹⁰ ܝܫܘܥ ܕܝܢ ܝܕܥ ܘܐܡܪ ܠܗܘܢ ܡܢܐ ܡܠܐܝܢ ܐܢܬܘܢ
ܠܗ ܠܐܢܬܬܐ ܥܒܕܐ ܫܦܝܪܐ ܥܒܕܬ ܠܘܬܝ ܀ ¹¹ ܒܟܠܙܒܢ ܓܝܪ ܡܣܟܢܐ ܐܝܬ
ܠܟܘܢ ܥܡܟܘܢ ܠܝ ܕܝܢ ܠܐ ܒܟܠܙܒܢ ܐܝܬ ܠܟܘܢ ܀ ¹² ܗܕܐ ܕܝܢ ܕܐܪܡܝܬ
ܒܣܡܐ ܗܢܐ ܥܠ ܦܓܪܝ ܐܝܟ ܕܠܡܩܒܪܢܝ ܥܒܕܬ ܀ ¹³ ܘܐܡܝܢ ܐܡܪ ܐܢܐ
ܠܟܘܢ ܕܐܝܟܐ ܕܬܬܟܪܙ ܣܒܪܬܝ ܗܕܐ ܒܟܠܗ ܥܠܡܐ ܢܬܡܠܠ ܐܦ ܡܕܡ
ܕܥܒܕܬ ܗܕܐ ܠܥܘܗܕܢܗ ܀ ¹⁴ ܗܝܕܝܢ ܐܙܠ ܚܕ ܡܢ ܬܪܥܣܪ ܕܡܬܩܪܐ ܝܗܘܕܐ

ܣܟܪܝܘܛܐ ܠܘܬ ܪ̈ܒܝ ܟܗܢ̈ܐ. 15 ܘܐܡܪ ܠܗܘܢ ܡܢܐ ܨܒܝܢ ܐܢܬܘܢ ܠܡܬܠ ܠܝ ܘܐܢܐ ܡܫܠܡ ܐܢܐ ܠܗ ܠܟܘܢ ܗܢܘܢ ܕܝܢ ܐܩܝܡܘ ܠܗ ܬܠܬܝܢ ܕܟܣܦܐ. 16 ܘܡܢ ܗܝܕܝܢ ܒܥܐ ܗܘܐ ܠܗ ܦܠܚܢܐ ܕܢܫܠܡܝܘܗܝ. 17 ܒܝܘܡܐ ܕܝܢ ܩܕܡܝܐ ܕܦܛܝܪ̈ܐ ܩܪܒܘ ܬܠܡܝܕ̈ܐ ܠܘܬ ܝܫܘܥ ܘܐܡܪܝܢ ܠܗ ܐܝܟܐ ܨܒܐ ܐܢܬ ܕܢܛܝܒ ܠܟ ܕܬܐܟܘܠ ܦܨܚܐ. 18 ܗܘ ܕܝܢ ܐܡܪ ܠܗܘܢ ܙܠܘ ܠܡܕܝܢܬܐ ܠܘܬ ܦܠܢ ܘܐܡܪܘ ܠܗ ܪܒܢ ܐܡܪ ܙܒܢܝ ܡܛܐ ܠܗ ܠܘܬܟ ܥܒܕ ܐܢܐ ܦܨܚܐ ܥܡ ܬܠܡܝܕ̈ܝ. 19 ܘܬܠܡܝܕ̈ܘܗܝ ܥܒܕܘ ܐܝܟܢܐ ܕܦܩܕ ܠܗܘܢ ܝܫܘܥ ܘܛܝܒܘ ܦܨܚܐ. 20 ܘܟܕ ܗܘܐ ܪܡܫܐ ܣܡܝܟ ܗܘܐ ܥܡ ܬܪܥܣܪ ܬܠܡܝܕ̈ܘܗܝ. 21 ܘܟܕ ܠܥܣܝܢ ܐܡܪ. ܐܡܝܢ ܐܡܪ ܐܢܐ ܠܟܘܢ ܕܚܕ ܡܢܟܘܢ ܡܫܠܡ ܠܝ. 22 ܘܟܪܝܬ ܠܗܘܢ ܛܒ ܘܫܪܝܘ ܠܡܐܡܪ ܠܗ ܚܕ ܚܕ ܡܢܗܘܢ ܠܡܐ ܐܢܐ ܡܪܝ. 23 ܗܘ ܕܝܢ ܥܢܐ ܘܐܡܪ ܗܘ ܕܨܒܥ ܐܝܕܗ ܥܡܝ ܒܠܓܝܢܬܐ ܗܘ ܢܫܠܡܢܝ. 24 ܘܗܐ ܒܪܗ ܕܐܢܫܐ ܐܙܠ ܐܝܟܢܐ ܕܟܬܝܒ ܥܠܘܗܝ. ܘܝ ܠܗ ܕܝܢ ܠܓܒܪܐ ܗܘ ܕܒܐܝܕܗ ܡܫܬܠܡ ܒܪܗ ܕܐܢܫܐ ܦܩܚ ܗܘܐ ܠܗ ܠܓܒܪܐ ܗܘ ܕܠܐ ܐܬܝܠܕ. 25 ܥܢܐ ܝܗܘܕܐ ܡܫܠܡܢܐ ܘܐܡܪ ܕܠܡܐ ܐܢܐ ܗܘ ܪܒܝ. ܐܡܪ ܠܗ ܝܫܘܥ ܐܢܬ ܐܡܪܬ. 26 ܟܕ ܕܝܢ ܠܥܣܝܢ ܢܣܒ ܝܫܘܥ ܠܚܡܐ ܘܒܪܟ ܘܩܨܐ ܘܝܗܒ ܠܬܠܡܝܕ̈ܘܗܝ ܘܐܡܪ. ܣܒܘ ܐܟܘܠܘ. ܗܢܘ ܦܓܪܝ. 27 ܘܢܣܒ ܟܣܐ ܘܐܘܕܝ ܘܝܗܒ ܠܗܘܢ ܘܐܡܪ. ܣܒܘ ܐܫܬܘ ܡܢܗ ܟܠܟܘܢ. 28 ܗܢܘ ܕܡܝ ܕܕܝܬܩܐ ܚܕܬܐ ܕܚܠܦ ܣܓܝܐܐ ܡܬܐܫܕ ܠܫܘܒܩܢܐ ܕܚܛܗ̈ܐ. 29 ܐܡܪ ܐܢܐ ܠܟܘܢ ܕܝܢ ܕܠܐ ܐܫܬܐ ܡܢ ܗܫܐ

ܡܢ ܗܢܐ ܝܠܕܐ ܕܓܦܬܐ ܠܝܘܡܐ ܕܒܗ ܐܫܬܝܘܗܝ ܥܡܟܘܢ ܚܕܬܐ
ܒܡܠܟܘܬܗ ܕܐܒܝ܀ 30 ܘܫܒܚܘ ܘܢܦܩܘ ܠܛܘܪܐ ܕܙܝܬܐ܀ 31 ܗܝܕܝܢ ܐܡܪ
ܠܗܘܢ ܝܫܘܥ ܐܢܬܘܢ ܟܠܟܘܢ ܬܬܟܫܠܘܢ ܒܝ ܒܗܢܐ ܠܠܝܐ. ܟܬܝܒ ܓܝܪ
ܕܐܡܚܐ ܠܪܥܝܐ ܘܢܬܒܕܪܘܢ ܥܢܐ ܕܡܪܥܝܬܗ܀ 32 ܡܢ ܒܬܪ ܕܩܐܡ ܐܢܐ ܕܝܢ ܩܕܡ
ܐܢܐ ܠܟܘܢ ܠܓܠܝܠܐ܀ 33 ܥܢܐ ܟܐܦܐ ܘܐܡܪ ܠܗ ܐܦܢ ܟܠܗܘܢ ܢܬܟܫܠܘܢ ܒܟ
ܐܢܐ ܡܡܬܘܡ ܠܐ ܐܬܟܫܠ ܒܟ܀ 34 ܐܡܪ ܠܗ ܝܫܘܥ ܐܡܝܢ ܐܡܪ ܐܢܐ ܠܟ
ܕܒܗܢܐ ܠܠܝܐ ܩܕܡ ܕܢܩܪܐ ܬܪܢܓܠܐ ܬܠܬ ܙܒܢܝܢ ܬܟܦܘܪ ܒܝ܀ 35 ܐܡܪ
ܠܗ ܟܐܦܐ. ܐܢ ܢܗܘܐ ܠܝ ܥܡܟ ܠܡܡܬ ܠܐ ܐܟܦܘܪ ܒܟ. ܘܗܟܢܐ ܐܦ
ܟܠܗܘܢ ܬܠܡܝܕܐ ܐܡܪܘ܀ 36 ܗܝܕܝܢ ܐܬܐ ܥܡܗܘܢ ܝܫܘܥ ܠܕܘܟܬܐ
ܕܡܬܩܪܝܐ ܓܕܣܡܢܝ ܘܐܡܪ ܠܬܠܡܝܕܘܗܝ. ܬܒܘ ܗܪܟܐ ܥܕ ܐܙܠ ܐܨܠܐ܀
37 ܘܕܒܪ ܠܟܐܦܐ ܘܠܬܪܝܗܘܢ ܒܢܝ ܙܒܕܝ ܘܫܪܝ ܠܡܬܥܩܘ ܘܠܡܬܟܪܝܘ܀
38 ܘܐܡܪ ܠܗܘܢ ܟܪܝܐ ܗܝ ܠܗ ܠܢܦܫܝ ܥܕܡܐ ܠܡܘܬܐ. ܩܘܘ ܠܝ ܗܪܟܐ
ܘܫܗܪܘ ܥܡܝ܀ 39 ܘܩܪܒ ܩܠܝܠ ܘܢܦܠ ܥܠ ܐܦܘܗܝ ܘܡܨܠܐ ܗܘܐ ܘܐܡܪ.
ܐܒܝ ܐܢ ܡܫܟܚܐ ܢܥܒܪ ܡܢܝ ܟܣܐ ܗܢܐ. ܒܪܡ ܠܐ ܐܝܟ ܕܐܢܐ ܨܒܐ ܐܢܐ ܐܠܐ ܐܝܟ
ܕܐܢܬ܀ 40 ܘܐܬܐ ܠܘܬ ܬܠܡܝܕܘܗܝ ܘܐܫܟܚ ܐܢܘܢ ܟܕ ܕܡܟܝܢ ܘܐܡܪ
ܠܟܐܦܐ ܗܟܢܐ ܠܐ ܐܫܟܚܬܘܢ ܚܕܐ ܫܥܐ ܕܬܫܗܪܘܢ ܥܡܝ܀ 41 ܐܬܬܥܝܪܘ ܘܨܠܘ
ܕܠܐ ܬܥܠܘܢ ܒܢܣܝܘܢܐ. ܪܘܚܐ ܡܛܝܒܐ ܦܓܪܐ ܕܝܢ ܟܪܝܗ܀ 42 ܬܘܒ ܐܙܠ
ܕܬܪܬܝܢ ܨܠܝ ܘܐܡܪ ܐܒܝ ܐܢ ܠܐ ܡܫܟܚܐ ܗܢܐ ܟܣܐ ܕܢܥܒܪ ܐܠܐ ܐܢ
ܐܫܬܝܬܗ ܢܗܘܐ ܨܒܝܢܟ܀ 43 ܘܐܬܐ ܬܘܒ ܐܫܟܚ ܐܢܘܢ ܟܕ ܕܡܟܝܢ.

ܥܝܢܝܗܘܢ ܓܝܪ ܝܩܝܪܢ ܗܘܝ܀ 44 ܘܫܒܩ ܐܢܘܢ ܘܐܙܠ ܬܘܒ ܨܠܝ ܕܬܠܬ
ܙܒܢܝܢ. ܘܠܗ ܠܡܠܬܐ ܐܡܪ܀ 45 ܗܝܕܝܢ ܐܬܐ ܠܘܬ ܬܠܡܝܕܘܗܝ ܘܐܡܪ
ܠܗܘܢ. ܕܡܟܘ ܡܟܝܠ ܘܐܬܬܢܝܚܘ. ܗܐ ܡܛܝܬ ܫܥܬܐ ܘܒܪܗ ܕܐܢܫܐ
ܡܫܬܠܡ ܒܐܝܕܝܗܘܢ ܕܚܛܝܐ܀ 46 ܩܘܡܘ ܢܐܙܠ. ܗܐ ܡܛܐ ܗܘ ܕܡܫܠܡ
ܠܝ܀ 47 ܘܥܕ ܗܘ ܡܡܠܠ ܗܐ ܝܗܘܕܐ ܡܫܠܡܢܐ ܚܕ ܡܢ ܬܪܥܣܪܬܐ ܐܬܐ
ܘܥܡܗ ܟܢܫܐ ܣܓܝܐܐ. ܥܡ ܣܦܣܪܐ ܘܚܘܛܪܐ ܡܢ ܠܘܬ ܪܒܝ ܟܗܢܐ ܘܩܫܝܫܐ
ܕܥܡܐ܀ 48 ܘܝܗܒ ܗܘܐ ܠܗܘܢ ܐܬܐ ܝܗܘܕܐ ܡܫܠܡܢܐ ܘܐܡܪ. ܠܗܘ ܕܢܫܩ
ܐܢܐ ܗܘܝܘ ܠܗ ܐܚܘܕܘ܀ 49 ܘܡܚܕܐ ܩܪܒ ܠܘܬ ܝܫܘܥ ܘܐܡܪ. ܫܠܡ ܪܒܝ
ܘܢܫܩܗ܀ 50 ܗܘ ܕܝܢ ܝܫܘܥ ܐܡܪ ܠܗ ܥܠ ܗܝ ܕܐܬܝܬ ܚܒܪܝ. ܗܝܕܝܢ
ܐܬܩܪܒܘ ܘܐܪܡܝܘ ܐܝܕܝܗܘܢ ܥܠ ܝܫܘܥ ܘܐܚܕܘܗܝ܀ 51 ܘܗܐ ܚܕ ܡܢ ܗܠܝܢ
ܕܥܡ ܝܫܘܥ ܐܘܫܛ ܐܝܕܗ ܘܫܡܛ ܣܦܣܪܐ ܘܡܚܝܗܝ ܠܥܒܕܐ ܕܪܒ ܟܘܡܪܐ
ܘܫܩܠܗ ܐܕܢܗ܀ 52 ܗܝܕܝܢ ܐܡܪ ܠܗ ܝܫܘܥ ܐܗܦܟ ܣܦܣܪܐ ܠܕܘܟܬܗ.
ܟܠܗܘܢ ܓܝܪ ܗܠܝܢ ܕܢܣܒܘ ܣܦܣܪܐ ܒܣܦܣܪܐ ܢܡܘܬܘܢ܀ 53 ܐܘ ܣܒܪ ܐܢܬ ܕܠܐ
ܡܫܟܚ ܐܢܐ ܕܐܒܥܐ ܡܢ ܐܒܝ ܘܢܩܝܡ ܠܝ ܗܫܐ ܝܬܝܪ ܡܢ ܬܪܬܥܣܪ
ܠܓܝܘܢܝܢ ܕܡܠܐܟܐ܀ 54 ܐܝܟܢܐ ܗܟܝܠ ܢܬܡܠܝܢ ܟܬܒܐ ܕܗܟܢܐ ܘܠܐ ܕܢܗܘܐ܀
55 ܒܗ ܒܫܥܬܐ ܐܡܪ ܝܫܘܥ ܠܟܢܫܐ. ܐܝܟ ܕܥܠ ܓܝܣܐ ܢܦܩܬܘܢ ܒܣܦܣܪܐ
ܘܒܚܘܛܪܐ ܕܬܐܚܕܘܢܢܝ. ܟܠܝܘܡ ܠܘܬܟܘܢ ܒܗܝܟܠܐ ܝܬܒ ܗܘܝܬ ܘܡܠܦ
ܘܠܐ ܐܚܕܬܘܢܢܝ. ܗܕܐ ܕܝܢ ܕܗܘܬ ܕܢܬܡܠܘܢ ܟܬܒܐ ܕܢܒܝܐ܀ 56 ܗܝܕܝܢ
ܬܠܡܝܕܐ ܟܠܗܘܢ ܫܒܩܘܗܝ ܘܥܪܩܘ܀ 57 ܘܗܢܘܢ ܕܐܚܕܘܗܝ ܠܝܫܘܥ

ܐܘܒܠܘܗܝ ܠܘܬ ܩܝܦܐ ܪܒ ܟܗܢܐ ܐܝܟܐ ܕܣܦܪܐ ܘܩܫܝܫܐ ܟܢܝܫܝܢ ܗܘܘ ܀
58 ܫܡܥܘܢ ܕܝܢ ܟܐܦܐ ܐܙܠ ܗܘܐ ܒܬܪܗ ܡܢ ܪܘܚܩܐ ܥܕܡܐ ܠܕܪܬܗ ܕܪܒ
ܟܗܢܐ ܘܥܠ ܝܬܒ ܠܓܘ ܥܡ ܕܚܫܐ ܕܢܚܙܐ ܚܪܬܐ ܀ 59 ܪܒܝ ܟܗܢܐ ܕܝܢ
ܘܩܫܝܫܐ ܘܟܢܘܫܬܐ ܟܠܗ ܒܥܝܢ ܗܘܘ ܥܠ ܝܫܘܥ ܣܗܕܘܬܐ ܕܫܘܩܪܐ ܐܝܟ ܕܢܡܝܬܘܢܝܗܝ
ܘܠܐ ܐܫܟܚܘ ܀ 60 ܘܐܬܘ ܣܓܝܐܐ ܣܗܕܐ ܕܫܘܩܪܐ ܠܚܪܬܐ ܕܝܢ ܩܪܒܘ ܬܪܝܢ
ܘܐܡܪܝܢ ܀ 61 ܗܢܐ ܐܡܪ ܕܡܫܟܚ ܐܢܐ ܕܐܫܪܐ ܗܝܟܠܗ ܕܐܠܗܐ ܘܠܬܠܬܐ
ܝܘܡܝܢ ܐܒܢܝܘܗܝ ܀ 62 ܘܩܡ ܪܒ ܟܗܢܐ ܘܐܡܪ ܠܗ. ܠܐ ܡܦܢܐ ܐܢܬ
ܦܬܓܡܐ. ܡܢܐ ܡܣܗܕܝܢ ܥܠܝܟ ܗܠܝܢ ܀ 63 ܝܫܘܥ ܕܝܢ ܫܬܝܩ ܗܘܐ
ܘܥܢܐ ܪܒ ܟܗܢܐ ܘܐܡܪ ܠܗ. ܡܘܡܐ ܐܢܐ ܠܟ ܒܐܠܗܐ ܚܝܐ ܕܬܐܡܪ ܠܢ ܐܢ
ܐܢܬ ܗܘ ܡܫܝܚܐ ܒܪܗ ܕܐܠܗܐ ܀ 64 ܐܡܪ ܠܗ ܝܫܘܥ. ܐܢܬ ܐܡܪܬ. ܐܡܪ ܐܢܐ
ܠܟܘܢ ܕܡܢ ܗܫܐ ܬܚܙܘܢ ܠܒܪܗ ܕܐܢܫܐ ܕܝܬܒ ܡܢ ܝܡܝܢܐ ܕܚܝܠܐ
ܘܐܬܐ ܥܠ ܥܢܢܝ ܫܡܝܐ ܀ 65 ܗܝܕܝܢ ܪܒ ܟܗܢܐ ܨܪܐ ܢܚܬܘܗܝ ܘܐܡܪ. ܗܐ
ܓܕܦ. ܡܢܐ ܡܬܒܥܝܢ ܠܢ ܣܗܕܐ. ܗܐ ܗܫܐ ܫܡܥܬܘܢ ܓܘܕܦܗ ܀
66 ܡܢܐ ܨܒܝܢ ܐܢܬܘܢ. ܥܢܘ ܘܐܡܪܝܢ ܚܝܒ ܗܘ ܡܘܬܐ ܀ 67 ܗܝܕܝܢ ܪܩܘ
ܒܐܦܘܗܝ ܘܡܩܦܚܝܢ ܗܘܘ ܠܗ. ܐܚܪܢܐ ܕܝܢ ܡܚܝܢ ܗܘܘ ܠܗ ܀
68 ܘܐܡܪܝܢ ܐܬܢܒܐ ܠܢ ܡܫܝܚܐ ܡܢܘ ܗܘ ܕܡܚܟ ܀ 69 ܟܐܦܐ ܕܝܢ ܝܬܒ
ܗܘܐ ܠܒܪ ܒܕܪܬܐ ܘܩܪܒܬ ܠܘܬܗ ܐܡܬܐ ܚܕܐ ܘܐܡܪܐ ܠܗ. ܐܦ ܐܢܬ ܥܡ
ܝܫܘܥ ܗܘܝܬ ܢܨܪܝܐ ܀ 70 ܗܘ ܕܝܢ ܟܦܪ ܩܕܡ ܟܠܗܘܢ ܘܐܡܪ. ܠܐ ܝܕܥ ܐܢܐ
ܡܢܐ ܐܡܪܐ ܐܢܬܝ ܀ 71 ܘܟܕ ܢܦܩ ܠܬܪܥܐ ܚܙܬܗ ܐܚܪܬܐ ܘܐܡܪܐ ܠܗܘܢ.

ܕܬܡܢ ܗܘܘ ܐܦ ܗܢܐ ܥܡ ܝܫܘܥ ܢܨܪܝܐ ܀ ⁷² ܘܬܘܒ ܟܦܪ ܒܡܘܡܬܐ.
ܕܠܐ ܝܕܥ ܐܢܐ ܠܗ ܠܓܒܪܐ ܀ ⁷³ ܒܬܪ ܩܠܝܠ ܕܝܢ ܩܪܒܘ ܗܢܘܢ ܕܩܝܡܝܢ
ܘܐܡܪܘ ܠܟܐܦܐ. ܫܪܝܪܐܝܬ ܐܦ ܐܢܬ ܡܢܗܘܢ ܐܢܬ. ܐܦ ܡܡܠܠܟ ܓܝܪ
ܡܘܕܥ ܠܟ ܀ ⁷⁴ ܗܝܕܝܢ ܫܪܝ ܠܡܚܪܡܘ ܘܠܡܘܡܐ. ܕܠܐ ܝܕܥ ܐܢܐ ܠܗ
ܠܓܒܪܐ. ܘܒܗ ܒܫܥܬܐ ܩܪܐ ܬܪܢܓܠܐ ܀ ⁷⁵ ܘܐܬܕܟܪ ܟܐܦܐ ܡܠܬܗ ܕܝܫܘܥ
ܕܐܡܪ ܗܘܐ ܠܗ. ܕܥܕܠܐ ܢܩܪܐ ܬܪܢܓܠܐ ܬܠܬ ܙܒܢܝܢ ܬܟܦܘܪ ܒܝ. ܘܢܦܩ
ܠܒܪ ܒܟܐ ܡܪܝܪܐܝܬ ܀

Chapter XXVII.

¹ ܟܕ ܕܝܢ ܗܘܐ ܨܦܪܐ ܡܠܟܐ ܢܣܒܘ ܥܠܗ ܕܝܫܘܥ ܪ̈ܒܝ ܟܗܢܐ ܘܩܫ̈ܝܫܐ
ܕܥܡܐ ܐܝܟ ܕܢܡܝܬܘܢܝܗܝ ܀ ² ܘܐܣܪܘܗܝ ܘܐܘܒܠܘܗܝ ܘܐܫܠܡܘܗܝ
ܠܦܝܠܛܘܣ ܗܓܡܘܢܐ ܀ ³ ܗܝܕܝܢ ܝܗܘܕܐ ܡܫܠܡܢܐ ܟܕ ܚܙܐ ܕܐܬܚܝܒ
ܝܫܘܥ ܐܬܬܘܝ. ܘܐܙܠ ܐܗܦܟ ܗܠܝܢ ܬܠܬܝܢ ܕܟܣܦܐ ܠܪ̈ܒܝ ܟܗܢܐ
ܘܠܩܫ̈ܝܫܐ ܀ ⁴ ܘܐܡܪ ܚܛܝܬ ܕܐܫܠܡܬ ܕܡܐ ܙܟܝܐ. ܗܢܘܢ ܕܝܢ ܐܡܪܘ ܠܗ.
ܠܢ ܡܐ ܠܢ. ܐܢܬ ܝܕܥ ܐܢܬ ܀ ⁵ ܘܫܕܐ ܟܣܦܐ ܒܗܝܟܠܐ ܘܫܢܝ ܘܐܙܠ
ܚܢܩ ܢܦܫܗ ܀ ⁶ ܪ̈ܒܝ ܟܗܢܐ ܕܝܢ ܢܣܒܘܗܝ ܠܟܣܦܐ ܘܐܡܪܘ. ܠܐ ܫܠܝܛ
ܕܢܪܡܝܘܗܝ ܒܝܬ ܩܘܪܒܢܐ ܡܛܠ ܕܛܝܡܝ ܕܡܐ ܗܘ ܀ ⁷ ܘܡܠܟܐ ܢܣܒܘ
ܘܙܒܢܘ ܒܗ ܐܓܘܪܣܐ ܕܩܘܩܝܐ ܠܒܝܬ ܩܒܘܪܐ ܕܐܟܣ̈ܢܝܐ ܀ ⁸ ܡܛܠ ܗܢܐ
ܐܬܩܪܝ ܐܓܘܪܣܐ ܗܘ ܩܪܝܬܐ ܕܕܡܐ ܥܕܡܐ ܠܝܘܡܢܐ ܀ ⁹ ܗܝܕܝܢ ܐܬܡܠܝ

ܡܛܠ ܕܐܬܐܡܪ ܥܠ ܐܝܕ ܢܒܝܐ ܕܐܡܪ. ܕܫܩܠܘܗ ܠܬܠܬܝܢ ܕܟܣܦܐ ܕܛܝܡܗ ܕܡܝܩܪܐ

ܕܥܒܕܘ ܡܢ ܒܢܝ ܐܝܣܪܝܠ܀ ¹⁰ ܘܝܗܒܠܗ ܐܢܝ ܠܐܓܘܪܗ ܕܦܚܪܐ ܐܝܟ ܕܦܩܕ

ܠܝ ܡܪܝܐ܀ ¹¹ ܗܘ ܕܝܢ ܝܫܘܥ ܩܡ ܩܕܡ ܗܓܡܘܢܐ. ܘܡܫܐܠܠܗ ܗܓܡܘܢܐ

ܘܐܡܪ ܠܗ. ܐܢܬ ܗܘ ܡܠܟܐ ܕܝܗܘܕܝܐ. ܐܡܪ ܠܗ ܝܫܘܥ. ܐܢܬ ܐܡܪܬ܀

¹² ܘܟܕ ܐܟܠܝܢ ܗܘܘ ܩܪܨܘܗܝ ܪ̈ܒܝ ܟܗ̈ܢܐ ܘܩܫ̈ܝܫܐ. ܡܕܡ ܦܬܓܡܐ ܗܘ ܠܐ

ܦܠܛ܀ ¹³ ܗܝܕܝܢ ܐܡܪ ܠܗ ܦܝܠܛܘܣ. ܠܐ ܫܡܥܬ ܐܢܬ ܟܡܐ ܡܣܗܕܝܢ

ܥܠܝܟ܀ ¹⁴ ܘܠܐ ܦܢܝ ܠܗ ܦܬܓܡܐ ܐܦܠܐ ܚܕܐ ܡܠܐ. ܘܥܠ ܗܕܐ ܐܬܕܡܪ

ܡܢܗ܀ ¹⁵ ܒܥܐܕܐ ܕܝܢ ܥܝܕܐ ܗܘܐ ܗܓܡܘܢܐ ܕܫܪܐ ܐܣܝܪܐ ܚܕ ܠܟܢܫܐ

ܐܝܢܐ ܕܗܢܘܢ ܨܒܝܢ ܗܘܘ܀ ¹⁶ ܐܝܬ ܗܘܐ ܠܗܘܢ ܕܝܢ ܐܣܝܪܐ ܝܕܝܥܐ ܕܡܬܩܪܐ

ܒܪ ܐܒܐ܀ ¹⁷ ܘܟܕ ܟܢܝܫܝܢ ܐܡܪ ܠܗܘܢ ܦܝܠܛܘܣ ܠܡܢ ܨܒܝܢ ܐܢܬܘܢ

ܕܐܫܪܐ ܠܟܘܢ ܠܒܪ ܐܒܐ ܐܘ ܠܝܫܘܥ ܕܡܬܩܪܐ ܡܫܝܚܐ܀ ¹⁸ ܝܕܥ ܗܘܐ ܓܝܪ

ܦܝܠܛܘܣ ܕܡܢ ܚܣܡܐ ܐܫܠܡܘܗܝ܀ ¹⁹ ܟܕ ܝܬܒ ܕܝܢ ܗܓܡܘܢܐ ܥܠ

ܒܝܡ ܕܝܠܗ ܫܕܪܬ ܠܗ ܐܢܬܬܗ ܘܐܡܪܐ ܠܗ. ܠܐ ܠܟ ܘܠܗ ܙܕܝܩܐ.

ܣܓܝ ܓܝܪ ܚܫܬ ܒܚܠܡܐ ܝܘܡܢܐ ܡܛܠܬܗ܀ ²⁰ ܪ̈ܒܝ ܟܗ̈ܢܐ ܕܝܢ

ܘܩܫ̈ܝܫܐ ܐܦܝܣܘ ܠܟܢ̈ܫܐ ܕܢܫܐܠܘܢ ܠܒܪ ܐܒܐ. ܠܝܫܘܥ ܕܝܢ ܕܢܘܒܕܘܢ܀

²¹ ܘܥܢܐ ܗܓܡܘܢܐ ܘܐܡܪ ܠܗܘܢ. ܠܡܢ ܨܒܝܢ ܐܢܬܘܢ ܕܐܫܪܐ ܠܟܘܢ ܡܢ

ܬܪ̈ܝܗܘܢ. ܗܢܘܢ ܕܝܢ ܐܡܪܘ. ܠܒܪ ܐܒܐ܀ ²² ܐܡܪ ܠܗܘܢ ܦܝܠܛܘܣ.

ܘܠܝܫܘܥ ܕܡܬܩܪܐ ܡܫܝܚܐ ܡܢܐ ܐܥܒܕ ܠܗ܀ ²³ ܐܡܪܝܢ ܟܠܗܘܢ.

ܢܙܕܩܦ. ܐܡܪ ܠܗܘܢ ܦܝܠܛܘܣ. ܡܢܐ ܓܝܪ ܕܒܝܫ ܥܒܕ. ܗܢܘܢ ܕܝܢ

ܝܬܝܪܐܝܬ ܩܥܘ ܘܐܡܪܝܢ. ܢܙܕܩܦ ❖ 24 ܦܝܠܛܘܣ ܕܝܢ ܟܕ ܚܙܐ ܕܡܕܡ ܠܐ
ܡܘܬܪ ܐܠܐ ܝܬܝܪܐܝܬ ܫܓܫܐ ܗܘܐ ܢܣܒ ܡܝܐ ܐܫܝ ܐܝܕܘܗܝ ܠܥܝܢ ܥܡܐ
ܘܐܡܪ. ܡܚܣܝ ܐܢܐ ܡܢ ܕܡܗ ܕܗܢܐ ܙܕܝܩܐ. ܐܢܬܘܢ ܬܕܥܘܢ ❖ 25 ܘܥܢܘ ܟܠܗ
ܥܡܐ ܘܐܡܪܘ. ܕܡܗ ܥܠܝܢ ܘܥܠ ܒܢܝܢ ❖ 26 ܗܝܕܝܢ ܫܪܐ ܠܗܘܢ ܠܒܪ
ܐܒܐ. ܘܢܓܕ ܒܦܪܓܠܐ ܠܝܫܘܥ ܘܐܫܠܡܗ ܕܢܙܕܩܦ ❖ 27 ܗܝܕܝܢ ܐܣܛܪܛܝܘܛܐ
ܕܗܓܡܘܢܐ ܕܒܪܘܗܝ ܠܝܫܘܥ ܠܦܪܛܘܪܝܢ ܘܟܢܫܘ ܥܠܘܗܝ ܠܟܠܗ ܐܣܦܝܪ ❖
28 ܘܐܫܠܚܘܗܝ ܘܐܠܒܫܘܗܝ ܟܠܡܝܣ ܕܙܚܘܪܝܬܐ ❖ 29 ܘܓܕܠܘ ܟܠܝܠܐ
ܕܟܘܒܐ ܘܣܡܘ ܒܪܫܗ ܘܩܢܝܐ ܒܝܡܝܢܗ ܘܒܪܟܘ ܥܠ ܒܘܪܟܝܗܘܢ
ܩܕܡܘܗܝ ܘܡܒܙܚܝܢ ܗܘܘ ܒܗ ܘܐܡܪܝܢ ܫܠܡ ܠܟ ܡܠܟܐ ܕܝܗܘܕܝܐ ❖
30 ܘܪܩܘ ܒܦܪܨܘܦܗ ܘܢܣܒܘ ܩܢܝܐ ܘܡܚܝܢ ܗܘܘ ܒܗ ܥܠ ܪܫܗ ❖
31 ܘܟܕ ܒܙܚܘ ܒܗ ܐܫܠܚܘܗܝ ܟܠܡܝܣ ܘܐܠܒܫܘܗܝ ܢܚܬܘܗܝ
ܕܢܙܕܩܦ ❖ 32 ܘܟܕ ܢܦܩܝܢ ܐܫܟܚܘ ܓܒܪܐ ܩܘܪܝܢܝܐ ܕܫܡܗ ܫܡܥܘܢ ܠܗܢܐ
ܫܚܪܘ ܕܢܫܩܘܠ ܙܩܝܦܗ ❖ 33 ܘܐܬܘ ܠܕܘܟܬܐ ܕܡܬܩܪܝܐ ܓܓܘܠܬܐ ܗܝ
ܕܡܬܦܫܩܐ ܩܪܩܦܬܐ ❖ 34 ܘܝܗܒܘ ܠܗ ܕܢܫܬܐ ܚܠܐ ܕܡܚܠܛ ܒܡܪܪܬܐ.
ܘܛܥܡ ܘܠܐ ܨܒܐ ܠܡܫܬܐ ❖ 35 ܘܟܕ ܙܩܦܘܗܝ ܦܠܓܘ ܢܚܬܘܗܝ ܒܦܣܐ
ܕܢܬܡܠܐ ܡܕܡ ܕܐܬܐܡܪ ܒܢܒܝܐ ܕܦܠܓܘ ܢܚܬܝ ܠܗܘܢ ܘܥܠ ܠܒܘܫܝ
ܐܪܡܝܘ ܦܣܐ ❖ 36 ܘܝܬܒܝܢ ܗܘܘ ܘܢܛܪܝܢ ܠܗ ܬܡܢ ❖ 37 ܘܣܡܘ ܠܥܠ
ܡܢ ܪܫܗ ܥܠܬܐ ܕܡܘܬܗ ܒܟܬܒܐ. ܗܢܘ ܝܫܘܥ ܡܠܟܐ ܕܝܗܘܕܝܐ ❖ 38 ܘܐܙܕܩܦܘ
ܥܡܗ ܬܪܝܢ ܠܣܛܝܐ ܚܕ ܡܢ ܝܡܝܢܗ ܘܚܕ ܡܢ ܣܡܠܗ ❖ 39 ܐܝܠܝܢ ܕܝܢ

ܕܥܒܪܝܢ ܗܘܘ ܡܓܕܦܝܢ ܗܘܘ ܥܠܘܗܝ. ܘܡܢܝܕܝܢ ܪܫܝܗܘܢ ܀ [40] ܘܐܡܪܝܢ ܫܪܐ ܗܝܟܠܐ ܘܒܢܐ ܠܗ ܠܬܠܬܐ ܝܘܡܝܢ ܦܪܘܩ ܢܦܫܟ ܐܢ ܒܪܗ ܐܢܬ ܕܐܠܗܐ ܘܚܘܬ ܡܢ ܙܩܝܦܐ ܀ [41] ܗܟܘܬ ܐܦ ܪܒܝ ܟܗܢܐ ܡܒܙܚܝܢ ܗܘܘ ܥܡ ܣܦܪܐ ܘܩܫܝܫܐ ܘܦܪܝܫܐ ܘܐܡܪܝܢ ܀ [42] ܠܐܚܪܢܐ ܐܚܝ ܢܦܫܗ ܠܐ ܡܫܟܚ ܠܡܚܝܘ. ܐܢ ܡܠܟܗ ܗܘ ܕܝܣܪܐܝܠ ܢܚܘܬ ܗܫܐ ܡܢ ܙܩܝܦܐ ܕܢܚܙܐ ܘܢܗܝܡܢ ܒܗ ܀ [43] ܬܟܝܠ ܥܠ ܐܠܗܐ ܢܦܪܘܩܝܗܝ ܗܫܐ ܐܢ ܪܥܐ ܒܗ. ܐܡܪ ܓܝܪ ܕܒܪܗ ܐܢܐ ܕܐܠܗܐ ܀ [44] ܗܟܘܬ ܐܦ ܠܣܛܝܐ ܗܢܘܢ ܕܐܙܕܩܦܘ ܥܡܗ ܡܚܤܕܝܢ ܗܘܘ ܠܗ ܀ [45] ܡܢ ܫܬ ܫܥܝܢ ܕܝܢ ܗܘܐ ܚܫܘܟܐ ܥܠ ܟܠܗ ܐܪܥܐ ܥܕܡܐ ܠܫܥܝܢ ܬܫܥ ܀ [46] ܘܠܐܦܝ ܬܫܥ ܫܥܝܢ ܩܥܐ ܝܫܘܥ ܒܩܠܐ ܪܡܐ ܘܐܡܪ. ܐܝܠ ܐܝܠ ܠܡܢܐ ܫܒܩܬܢܝ ܀ [47] ܐܢܫܝܢ ܕܝܢ ܡܢ ܗܢܘܢ ܕܩܝܡܝܢ ܗܘܘ ܬܡܢ ܟܕ ܫܡܥܘ ܐܡܪܝܢ ܗܘܘ ܗܢܐ ܠܐܠܝܐ ܩܪܐ ܀ [48] ܘܒܗ ܒܫܥܬܐ ܪܗܛ ܚܕ ܡܢܗܘܢ ܘܢܣܒ ܐܣܦܘܓܐ ܘܡܠܝܗ ܚܠܐ ܘܣܡܗ ܒܩܢܝܐ ܘܡܫܩܐ ܗܘܐ ܠܗ ܀ [49] ܫܪܟܐ ܕܝܢ ܐܡܪܝܢ ܗܘܘ ܫܒܘܩ ܢܚܙܐ ܐܢ ܐܬܐ ܐܠܝܐ ܠܡܦܪܩܗ ܀ [50] ܗܘ ܕܝܢ ܝܫܘܥ ܬܘܒ ܩܥܐ ܒܩܠܐ ܪܡܐ ܘܫܒܩ ܪܘܚܗ ܀ [51] ܘܡܚܕܐ ܐܦܝ ܬܪܥܐ ܕܗܝܟܠܐ ܐܨܛܪܝ ܠܬܪܝܢ ܡܢ ܠܥܠ ܠܬܚܬ ܘܐܪܥܐ ܐܬܬܙܝܥܬ ܘܟܐܦܐ ܐܨܛܪܝ ܀ [52] ܘܩܒܪܐ ܐܬܦܬܚܘ ܘܓܘܫܡܐ ܣܓܝܐܐ ܕܩܕܝܫܐ ܕܫܟܝܒܝܢ ܗܘܘ ܩܡܘ ܀ [53] ܘܢܦܩܘ ܘܒܬܪ ܩܝܡܬܗ ܥܠܘ ܠܡܕܝܢܬܐ ܩܕܝܫܬܐ ܘܐܬܚܙܝܘ ܠܣܓܝܐܐ ܀ [54] ܩܢܛܪܘܢܐ ܕܝܢ ܘܐܝܠܝܢ ܕܥܡܗ ܕܢܛܪܝܢ ܗܘܘ ܠܝܫܘܥ ܟܕ ܚܙܘ ܙܘܥܐ ܘܐܝܠܝܢ ܕܗܘܝ ܕܚܠܘ ܛܒ ܘܐܡܪܘ. ܫܪܝܪܐܝܬ ܗܢܐ

ܒܪܗ ܗܘܐ ܕܐܠܗܐ ܀ ⁵⁵ ܐܝܬ ܗܘܝ̈ ܕܝܢ ܬܡܢ ܢܫ̈ܐ ܣܓܝܐܬܐ ܕܚܙܝܢ ܗܘܝ̈ ܡܢ ܪܘܚܩܐ ܗܠܝܢ ܕܐܬܝ̈ ܗܘܝ̈ ܒܬܪܗ ܕܝܫܘܥ ܡܢ ܓܠܝܠܐ ܘܡܫܡܫܢ ܗܘܝ̈ ܠܗ ܀ ⁵⁶ ܕܚܕܐ ܡܢܗܝܢ ܡܪܝܡ ܡܓܕܠܝܬܐ ܘܡܪܝܡ ܐܡܗ ܕܝܥܩܘܒ ܘܕܝܘܣܐ ܘܐܡܗܘܢ ܕܒ̈ܢܝ ܙܒܕܝ ܀ ⁵⁷ ܟܕ ܗܘܐ ܕܝܢ ܪܡܫܐ ܐܬܐ ܓܒܪܐ ܥܬܝܪܐ ܡܢ ܪܡܬܐ ܕܫܡܗ ܝܘܣܦ ܕܐܦ ܗܘ ܐܬܬܠܡܕ ܗܘܐ ܠܝܫܘܥ ܀ ⁵⁸ ܗܢܐ ܩܪܒ ܠܘܬ ܦܝܠܛܘܣ ܘܫܐܠ ܦܓܪܗ ܕܝܫܘܥ ܘܦܩܕ ܦܝܠܛܘܣ ܕܢܬܝܗܒ ܠܗ ܦܓܪܐ ܀ ⁵⁹ ܘܢܣܒܗ ܝܘܣܦ ܠܦܓܪܐ ܘܟܪܟܗ ܒܟܬܢܐ ܕܟܝܐ ܢܩܕܐ ܀ ⁶⁰ ܘܣܡܗ ܒܩܒܪܐ ܚܕܬܐ ܕܝܠܗ ܕܢܩܝܪ ܒܟܐܦܐ ܘܥܓܠܘ ܟܐܦܐ ܪܒܬܐ. ܐܪܡܝܘ ܥܠ ܬܪܥܐ ܕܒܝܬ ܩܒܘܪܐ ܘܐܙܠ ܀ ⁶¹ ܐܝܬ ܗܘܝ̈ ܕܝܢ ܬܡܢ ܡܪܝܡ ܡܓܕܠܝܬܐ ܘܡܪܝܡ ܐܚܪܬܐ ܕܝܬܒ̈ܢ ܗܘܝ̈ ܠܘܩܒܠ ܩܒܪܐ ܀ ⁶² ܠܝܘܡܐ ܕܝܢ ܕܒܬܪ ܥܪܘܒܬܐ ܐܬܟܢܫܘ ܪ̈ܒܝ ܟܗ̈ܢܐ ܘܦܪ̈ܝܫܐ ܠܘܬ ܦܝܠܛܘܣ ܀ ⁶³ ܘܐܡܪܝܢ ܠܗ. ܡܪܢ ܐܬܕܟܪܢ ܕܗܘ ܡܛܥܝܢܐ ܐܡܪ ܗܘܐ ܟܕ ܚܝ ܕܡܢ ܒܬܪ ܬܠܬܐ ܝܘܡܝܢ ܩܐܡ ܐܢܐ ܀ ⁶⁴ ܦܩܘܕ ܗܟܝܠ ܢܬܢܛܪ ܩܒܪܐ ܥܕܡܐ ܠܬܠܬܐ ܝܘܡܝܢ ܕܠܡܐ ܢܐܬܘܢ ܬܠܡܝܕܘܗܝ ܢܓܢܒܘܢܝܗܝ ܒܠܠܝܐ ܘܢܐܡܪܘܢ ܠܥܡܐ ܕܩܡ ܡܢ ܒܝܬ ܡܝ̈ܬܐ ܘܬܗܘܐ ܛܥܝܘܬܐ ܐܚܪܝܬܐ ܒܝܫܐ ܡܢ ܩܕܡܝܬܐ ܀ ⁶⁵ ܐܡܪ ܠܗܘܢ ܦܝܠܛܘܣ. ܐܝܬ ܠܟܘܢ ܩܣܛܘܢܪ̈ܐ. ܙܠܘ ܐܙܕܗܪܘ ܐܝܟ ܕܝܕܥܝܢ ܐܢܬܘܢ ܀ ⁶⁶ ܗܢܘܢ ܕܝܢ ܐܙܠܘ ܐܙܕܗܪܘ ܒܩܒܪܐ ܘܚܬܡܘ ܟܐܦܐ ܥܡ ܩܣܛܘܢܪ̈ܐ ܀

Chapter XXVIII.

[1] ܒܪܡܫܐ ܕܝܢ ܒܫܒܬܐ ܕܢܓܗ ܚܕ ܒܫܒܐ ܐܬܬ ܡܪܝܡ ܡܓܕܠܝܬܐ
ܘܡܪܝܡ ܐܚܪܬܐ ܕܢܚܙܝܢ ܩܒܪܐ ܀ [2] ܘܗܐ ܙܘܥܐ ܪܒܐ ܗܘܐ. ܡܠܐܟܐ ܓܝܪ ܕܡܪܝܐ
ܢܚܬ ܡܢ ܫܡܝܐ ܘܩܪܒ ܥܓܠ ܟܐܦܐ ܡܢ ܬܪܥܐ ܘܝܬܒ ܗܘܐ ܠܥܠܝܗ ܀
[3] ܐܝܬܘܗܝ ܗܘܐ ܕܝܢ ܚܙܘܗ ܐܝܟ ܒܪܩܐ ܘܠܒܘܫܗ ܚܘܪ ܗܘܐ ܐܝܟ ܬܠܓܐ ܀
[4] ܘܡܢ ܕܚܠܬܗ ܐܙܕܘܥܘ ܐܝܠܝܢ ܕܢܛܪܝܢ ܗܘܘ. ܘܗܘܘ ܐܝܟ ܡܝܬܐ ܀
[5] ܥܢܐ ܕܝܢ ܡܠܐܟܐ ܘܐܡܪ ܠܢܫܐ. ܐܢܬܝܢ ܠܐ ܬܕܚܠܝܢ. ܝܕܥ ܐܢܐ ܓܝܪ
ܕܠܝܫܘܥ ܕܐܙܕܩܦ ܒܥܝܢ ܐܢܬܝܢ ܀ [6] ܠܐ ܗܘܐ ܬܢܢ. ܩܡ ܠܗ ܓܝܪ ܐܝܟܢܐ
ܕܐܡܪ. ܬܝܝܢ ܚܙܝܝܢ ܕܘܟܬܐ ܕܣܝܡ ܗܘܐ ܒܗ ܡܪܢ ܀ [7] ܘܐܙܠܝܢ ܒܥܓܠ
ܘܐܡܪܝܢ ܠܬܠܡܝܕܘܗܝ ܕܩܡ ܡܢ ܒܝܬ ܡܝܬܐ ܘܗܐ ܩܕܡ ܠܟܘܢ ܠܓܠܝܠܐ
ܬܡܢ ܬܚܙܘܢܢܗ. ܗܐ ܐܡܪܬ ܠܟܝܢ ܀ [8] ܘܐܙܠ ܒܥܓܠ ܡܢ ܩܒܪܐ
ܒܕܚܠܬܐ ܘܒܚܕܘܬܐ ܪܒܬܐ ܘܪܗܛܝ ܠܡܐܡܪ ܠܬܠܡܝܕܘܗܝ ܀ [9] ܘܗܐ ܝܫܘܥ
ܦܓܥ ܒܗܝܢ ܘܐܡܪ ܠܗܝܢ. ܫܠܡ ܠܟܝܢ. ܗܢܝܢ ܕܝܢ ܩܪܒ ܐܚܕ
ܪܓܠܘܗܝ ܘܣܓܕ ܠܗ ܀ [10] ܗܝܕܝܢ ܐܡܪ ܠܗܝܢ ܝܫܘܥ ܠܐ ܬܕܚܠܝܢ ܐܠܐ
ܐܙܠܝܢ ܐܡܪܝܢ ܠܐܚܝ ܕܢܐܙܠܘܢ ܠܓܠܝܠܐ ܘܬܡܢ ܢܚܙܘܢܢܝ ܀ [11] ܟܕ ܐܙܠܝܢ
ܕܝܢ ܐܬܘ ܐܢܫܐ ܡܢ ܩܘܣܛܘܢܪܐ ܗܢܘܢ ܠܡܕܝܢܬܐ ܘܐܡܪܘ ܠܪܒܝ ܟܗܢܐ ܟܠ
ܡܕܡ ܕܗܘܐ ܀ [12] ܘܐܬܟܢܫܘ ܥܡ ܩܫܝܫܐ ܘܢܣܒܘ ܡܠܟܐ ܘܝܗܒܘ ܟܣܦܐ ܠܐ
ܙܥܘܪ ܠܩܣܛܘܢܪܐ ܀ [13] ܘܐܡܪܝܢ ܠܗܘܢ. ܐܡܪܘ ܕܬܠܡܝܕܘܗܝ ܐܬܘ
ܓܢܒܘܗܝ ܒܠܠܝܐ ܟܕ ܕܡܟܝܢܢ ܀ [14] ܘܐܢ ܐܫܬܡܥܬ ܗܕܐ ܩܕܡ ܗܓܡܘܢܐ
ܚܢܢ ܡܦܝܣܝܢܢ ܠܗ ܘܠܟܘܢ ܕܠܐ ܨܦܬܐ ܥܒܕܝܢܢ ܀ [15] ܗܢܘܢ ܕܝܢ ܟܕ ܢܣܒܘ
ܟܣܦܐ ܥܒܕܘ ܐܝܟ ܕܐܠܦܘ ܐܢܘܢ. ܘܢܦܩܬ ܡܠܬܐ ܗܕܐ ܒܝܬ ܝܗܘܕܝܐ ܥܕܡܐ
ܠܝܘܡܢܐ ܀ [16] ܬܠܡܝܕܐ ܕܝܢ ܚܕܥܣܪ ܐܙܠܘ ܠܓܠܝܠܐ ܠܛܘܪܐ ܐܝܟܐ ܕܩܒܥ
ܠܗܘܢ ܝܫܘܥ ܀ [17] ܘܟܕ ܚܙܐܘܗܝ ܣܓܕܘ ܠܗ. ܡܢܗܘܢ ܕܝܢ ܐܬܦܠܓܘ ܗܘܘ ܀
[18] ܘܩܪܒ ܝܫܘܥ ܡܠܠ ܥܡܗܘܢ ܘܐܡܪ ܠܗܘܢ. ܐܬܝܗܒ ܠܝ ܟܠ

ܡܢ ܟܠܗܝܢ ܒܫܡܝܐ ܘܒܐܪܥܐ. ܘܐܡܪܬܝ ܕܡܛܠ ܐܝܟ ܐܦ ܐܢܐ ܡܫܡܪ ܐܢܐ
ܠܟܘܢ ܀ 19 ܐܢܬ ܗܘ ܒܠܚܘܕ ܠܟܠܗܘܢ ܟܠܗܘܢ ܒܫܡܝܐ ܘܐܠܗܝܢ ܐܠܗܝ ܒܡܫܬ ܐܝܬ
ܘܒܢܝܐ ܘܢܘܪܐ ܕܡܕܒܪܐ ܀ 20 ܘܟܠܗܘܢ ܐܠܗܝ ܕܠܗܘܢ ܥܠ ܗܢܐ ܕܡܫܟܚܘܢ. ܘܗܘܐ
ܐܢܐ ܒܟܠܗܘܢ ܐܢܐ ܟܠܗܘܢ ܡܢܗܘܢ ܒܪܝܬܐ ܒܫܘܠܡܗ ܕܟܠܗܝܢ ܐܡܝܢ ܀

SELECTION FROM THE HISTORY OF RABBAN SOMA.

ܦܪܝܫܬ ܫܪܒܗ ܕܪܒܢ ܨܘܡܐ.

ܐܝܬ ܗܘܐ ܡܢ ܐܢܫ ܡܗܝܡܢܐ ܒܐܬܪܐ ܕܡܕܝܢܬ ܟܢܒܠܝܩ: ܒܢܝܐ ܡܢ ܒܢܝܢ ܕܟܠܒܐ ܘܒܢܝܢ ܕܓܢܣܐ: ܕܝܕܥܐ ܕܝܢ ܒܫܘܒܚܐ ܘܒܥܘܬܪܐ: ܕܫܡܗ ܫܐܢ ܦܘܪܐ. ܟܕ ܗܘܐ ܕܝܢ ܒܡܕܝܢܬܐ ܕܡܬܩܪܝܐ ܚܢ ܒܠܝܩ. ܐܘܠܕ ܒܕܒܝܬ ܒܫܘܠܡܐ ܕܡܠܟܘܬܐ ܕܡܘܓܠܝܐ. ܐܫܬܘܬܦ ܡܢ ܢܚܫܒܬ ܠܬܘܕܝܬܐ ܕܫܡܗ ܩܝܡܐ. ܘܟܕ ܢܚܬ ܠܗܘܢ ܒܪܐ ܘܠܐ ܗܘܐ ܠܗܘܢ ܡܘܬܪܐ: ܐܬܬܟܝܠܘ ܒܬܚܒܬܐ ܘܒܬܘܚܢܬܐ ܠܘܬ ܐܠܗܐ:

ܐܡܝܢܐ ܕܠܐ ܢܬܠܐ ܐܝܟ ܒܪ ܡܫܒܚܐ ܕܚܝܗ. ܘܡܫܒܚܐ. ܗܘ ܓܝܪ ܒܛܝܒܘܬܗ ܘܒܪܚܡܘܗܝ: ܦܬܚ ܬܪܥܗܘܢ. ܘܐܬܪܚܡ ܒܠܗܘܢ. ܡܛܠ ܗܢܐ ܓܝܪ ܠܡܫܠܡܘ ܦܘܩܕܢܗ ܕܝܠܗ. ܘܐܝܬܝܗܒ ܒܪܐ

ܕܡܒܣܡܢܐ ܘܡܟܢܫܐ. ܟܠ ܠܡ ܕܫܦܠ ܢܦܩ. ܘܕܒܪܐ
ܡܥܒܕ. ܘܠܐܡܢܐ ܕܢܦܩ ܡܬܦܬܚ ܠܗ. ܡܫܡܠܐܝܬ ܥܠ
ܦܓܪܐ ܒܪܝܬܐ ܝܒܝܫ. ܘܗܘܐ ܟܕ ܬܪܥܣܗ ܠܡܒܝܬܐ
ܡܫܬܒܚܢܐ ܕܒܒܪܐ ܥܠܬ ܘܒܝܫܬܐ: ܟܕ ܒܣܘܥܒܐ
ܬܘܒܝܐ ܒܣܓܝܐܬܐ ܡܦܪܓܝܢ. ܗܘܐ ܟܕ ܒܢܝܫܐ ܒܫܬܘܬܗ
ܕܫܠܡܢܐ ܠܐ ܡܬܦܠܝܢ: ܟܕ ܒܝܒܫܐ ܬܘܒܝܐ ܒܟܝ.
ܘܐܬܘܬܗ ܕܒܠܘܣ ܠܐ ܡܬܦܣܩ. ܟܕ ܠܡܬܠܐܒܐ
ܠܡܬܚܒܬ ܒܟܗ ܦܪܠܘܢܗ ܦܚܠܝܬ. ܐܝܟ ܕܝܢ ܕܘܫܐ
ܕܟܠܢܐ ܠܐܬܘܬܐ: ܘܠܐܪܝܬ ܒܪܐ. ܘܡܢܘ ܕܝܢ ܥܒܕܗ
ܝܘܟܐ. ܘܣܕܝܗ ܢܕܘܬܐ ܪܟܘܬܐ. ܘܢܕܝܗ ܒܟܠܕܗ
ܠܥܒܕܐ ܘܠܒܢܝ ܠܗܘܟܐ. ܘܟܕ ܒܬܘܪܒܬܐ ܡܦܠܣܬܐ
ܠܡܥܘܣܬܐ ܕܗܘܟܠ ܗܠܦܢܐ ܟܘܒܗ: ܠܟܠܦܢܐ ܕܥܒܐ
ܐܥܠܝܗܘܢ. ܘܡܢܟܘܢ ܣܦܢܐܝܬ ܒܗܠܦܝܢ ܒܕܘܬܢܐ
ܕܥܘܗܝ. ܘܐܡܟܪܘܢ ܘܣܕܝܗ ܒܗ. ܘܠܡܘܟܠ ܕܪܟ ܟܘܦܐ
ܢܥܬܘܢ. ܘܟܪܢܝܐ ܡܠܝܦܝܢܐ ܢܬܟܚܒ. ܗܘܝܐ ܕܝܢ
ܦܢܝܢܐ ܒܕܘܬܐ ܕܡܒܝܢܘܬܐ ܒܗܕܘܬܐ ܘܢܬܬܒܪ ܒܟܠܗ
ܢܟܘܬܐ ܘܡܟܒܟܘܬܐ. ܘܢܬܬܢܦܠ ܒܝܢܝ ܡܢܬܬܘܬܐ.
ܘܢܬܦܬܥ ܥܠ ܕܘܟܢܐ ܕܡܬܒܪܬܘܬܐ: ܟܕ ܡܠܒ ܠܗ ܝܫܪܢ
ܥܬܝܢ: ܘܢܬܒܪܥܝܬ ܒܠܟܗ ܢܘܪܐ ܐܠܗܘܬܐ. ܘܐܘܣܦܬ
ܠܒܬܪ ܣܠܝܩܘܬܐ ܘܢܦܢܬ ܠܢܦܫܗ ܕܗܘܬܐ ܡܢ ܝܘܡܢܐ

ܘܥܠ ܪ̈ܚܘܒܘܬܐ. ܘܝܫܡ ܠܒܪ ܢܛܪ ܡܢ ܥܠ ܡܕܡ ܗܘ̈ܝܐ
ܕܡ̈ܕܢܚܐ. ܘܠܐ ܝܕܥܐ ܕܟܒܪ ܢܘܟܪܝܐ ܕܦܛܪ̈ܝܐ ܕܐܝܬܘܗܝ ܢܫܘܪ
ܠܡܫܬܪܗ. ܠܡܠܟܘܬܐ ܕܥܠܡܐ ܥܒܪܐ ܒܫܢܬܐ. ܘܒܬܪ ܟܢܫܬܗ
ܚܕܪ ܡܫܢܬܐ. ܠܡܐܟܠܐ ܘܗܘܡܟܐ ܐܡܪ ܗܘ ܕܠܐ ܐܝܬܘܗܝ
ܠܗܘܢܐ ܫܪܒ. ܘܠܡܫܬܐ ܡܪܘܡܝܐ ܠܟܪܟܪ ܐܡܠܟ. ܟܕ
ܕܝܢ ܐܬܟܫܦ ܐܟܣܢܝܐ ܒܗܕܐ: ܢܦܩܐ ܪ̈ܟܐ ܐܪܙܝܢ ܐܝܟ.
ܘܡܐܟܐ ܒܡܟܬܒܐ ܢܝܦ ܠܗܘܢ: ܕܢܝܚܗܘܢ ܡܬܦܪܫ
ܡܢܗܘܢ. ܘܡܢ ܗܘܘ ܟܪ ܟܠܟܐ ܥܝܦܐ ܘܐܬܟܫܦܘ
ܠܗ. ܘܫܘܐܠܐ ܟܠܡ̈ܢܐ ܦܪܟܘ ܠܗ. ܠܡܢܐ: ܠܡ ܟܪܢ
ܢܨܒܬܐ: ܘܐܡܪ ܪܒܢ ܨܘܡܐ: ܐܡܢܐ ܘܡܒܥ ܐܠܗܐ
ܠܟܡܪ̈ܢ: ܠܡܢܐ: ܕܝܢ ܫܠܝܚ ܪܒܢ ܐܝܟܢ: ܘܢܕ ܕܠܡܢ ܦܐܦܐ
ܡܢܢ: ܐܬܡܢܝܢ ܕܐܡܝܢ ܡܫܝܚܐ: ܐܝܟܘܬܟܠ ܕܟܪܝܗ ܡܦܪܢܐ
ܠܡܝܢ: ܐܡܪܢܐ: ܟܡܢ ܠܝ ܢܬܚܙܐ ܕܪܡܢ ܘܫܟܢ: ܠܡܢ
ܡܪܝܫ ܘܫܟܢܝ ܕܢܟܪ̈ܝܐ ܢܪܐܝܢ ܠܝ. ܘܡܢ ܟܕܡ̈ܟܐ
ܟܪܐܝܢ ܗܠܝܢ ܐܦܝܫܘܗܝ: ܘܡܟܪܟܬܗܘܢ ܒܫܘܪܐ ܪ̈ܘܚܢܐ
ܐܟܘܫܗܘܢ. ܐܝܫܬܟܚ ܠܗܘܢ ܟܪ̈ܢܐܝܬ. ܘܪܒܢ ܟܡܟܗܘܢ
ܦܟܪ̈ܢܐܝܬ. ܠܐ ܕܝܢ ܝܟܢ̈ܐܝܬ. ܘܟܕ ܐܠܗ ܫܢܝܢ ܓܡܝܪ
ܐܝܢܘܢ ܠܐܟܫ̈ܘܐ ܦܟܪ̈ܢܝܐ: ܐܘܕܝ ܡܢ ܟܡܠܬܗ ܠܐ ܦܫ.
ܘܟܪܗܠܗ ܒܡܟܠܐ ܐܡܟܢܐܝܬ ܐܝܬܟܬܫ. ܟܕ ܫܡܘ ܕܝܢ
ܕܠܐ ܟܡܝܢܐ ܟܪܐܬܢܝܬܗܘܢ: ܘܐܡܪܝܢ ܠܐ ܡܕܡ

ܒܝܘܡܬܐ ܕܣܘܦܐ ܡܥܒܕܢܐ ܣܥܪܬܐ ܕܐܠܗܘܬܗ : ܠܝܘܚܢܢ
ܓܒܪܐ ܐܦ : ܡܢܝܢܗܘܢ ܟܠܗ ܕܝܢ ܗܘ. ܢܫܡܥܘܗܝ ܢܬܟܠܘܢ
ܠܚܘܒܗ ܘܡܟܝܟܘܬܗ. ܦܠܓ ܠܐܦܣܘܣ. ܘܦܩܕ ܠܐܦܝܣܩܘܦܐ
ܕܢܬܩܢ ܥܕܬܐ. ܘܐܬܦܪܫ ܡܢ ܬܡܢ ܒܪܒܥܐ: ܒܫܒܐ
ܡܢܝ. ܒܬܪܟܢ ܒܡܛܪܦܘܠܝܛܐ: ܘܚܙܐ ܠܝܘܚܢܢ ܒܟܢܘܫܬܐ
ܕܡܕܝܢܬܗ ܥܠ ܫܒܪܐ ܕܐܠܗܘܬܐ ܕܛܒܝܒܘܬܐ. ܘܗܘܝܢܐ
ܕܡܘܕܐ ܠܗܘܢ ܥܡܡܐ. ܘܕܝܫ ܐܢܫܐ ܕܚܙܘ ܡܫܠܡܢܘܬܐ.
ܦܪܫ ܠܗ ܡܠܝܘܬܐ ܘܫܒܩ ܟܠܗ ܢܦܫܗ ܥܡܗ ܥܒܕܐ. ܘܡܛܪܦ
ܝܬܝܪ ܕܢܬܒܥ ܡܢ ܒܢܝܢܫܐ ܘܢܬܠ ܢܦܫܗ ܠܡܠܟܐ
ܒܬܘܟܬܐ ܣܓܝܐ ܕܦܘܪܫܢܐ: ܐܝܟܢܐ ܕܟܠ ܢܬܚܝܒ
ܒܫܒܘܩܢܐ. ܢܟܘ ܕܝܢ ܘܐܢܠ ܐܓܘܣܐ ܕܐܢܫܐ ܚܕ ܕܡܢ
ܡܕܝܢܬܐ ܕܐܦܣܘܣ: ܘܟܕ ܠܗ ܠܝܘܚܢܢ ܗܘܐ. ܘܐܬܥܗܕ
ܕܘܟܝܬܐ ܚܕܐ ܕܐܝܬ ܒܗ ܡܕܝܢܬܐ. ܘܟܠ ܝܘܡ ܡܬܒܥܐ
ܕܢܬܬܠ ܠܡܠܟܐ ܗܘ. ܘܟܕ ܐܫܟܚ ܒܫܐܝܪ. ܘܦܩܕ ܠܒܢܝ
ܡܪܗ. ܗܘ ܕܠܐܢܫ ܗܠܝܢ ܐܬܥܝܪܘ ܠܡܕܝܢܬܐ ܘܢܦܩ ܠܒܪܐ
ܒܐܪܥܐ ܗܘ. ܘܥܕܝܘ ܡܚܟܝܡܝܢ ܐܢܫܐ ܕܐܪܘܗܝ. ܠܝܘܚܢܢ
ܡܠܟܘܗܝ. ܘܐܝܟܢܐ ܡܬܩܪܝܢ ܗܘܐ ܠܗ ܡܢ ܟܠܗ:

GLOSSARY.

ܐ.

ܐܒܕ to perish.

ܐܒܐ father, § 87. 1.

ܐܒܕܢܐ perdition.

ܐܒܠ grief.

ܐܓܝܪܐ hired.

ܐܓܪܐ hire.

ܐܓܘܪܣܐ field.

ܐܕܘܡܝܐ Edomite.

ܐܕܡ Adam.

ܐܕܡܬܐ ground.

ܐܕܢܐ ear.

ܐܘ or.

ܐܘܡܢ artificer.

ܐܘܝ oh!

ܐܘܪܚܐ way.

ܐܘܨܪܐ treasury.

ܐܘܪܫܠܡ Jerusalem.

ܐܙܠ to go, § 64. 1.

ܐܚܐ brother, § 87. 1.

ܐܚܪܝܢ afterwards.

ܐܚܪܝܬܐ the last.

ܐܚܪܢܐ other, next.

ܐܚܪܬܐ other, § 87. 4.

ܐܚܕ to seize.

ܐܚܕܢܐ possession.

ܐܝܟ according to, like, § 89 B 1.

ܐܝܟ ܕ according as, so that.

ܐܝܟܐ where?

ܐܝܟܘ where is?

ܐܝܟܢܐ as.

ܐܝܠ God.

ܐܝܠܝܢ who, which, what? §§ 39. 103.

ܐܝܠܢܐ tree.

ܐܝܡܟܐ whence?

ܐܝܢܐ who, which, what? §§ 39. 103.

ܐܝܣܪܝܠ Israel.

ܐܝܬܐ *see* ܝܬܐ.

ܐܝܩܪܐ glory, honor.

ܐܝܬ there is, §§ 65, 128.
ܐܟܚܕܐ together, as one.
ܐܝܟ like, § 89 B. 1.
ܐܟܠ to eat.
ܐܟܣܢܝܐ stranger.
ܐܠܗܐ God.
ܐܠܗܝܐ divine.
ܐܠܐ if not, unless, but.
ܐܠܘ if, § 138. 5.
ܐܠܝܐ Elijah.
ܐܠܥܐ rib.
ܐܠܦ to learn.
ܐܠܦ to teach.
ܐܠܦܐ ship.
ܐܡܐ mother, § 87. 7.
ܐܡܘܬܐ nations, § 86. 3; 87.3.
ܐܡܝܢ verily, amen.
ܐܡܝܢܐܝܬ always, ceaselessly.
ܐܡܪ to say.
ܐܡܬܐ maid.
ܐܡܬܝ when.
ܐܡܢ II to persevere.
ܐܢ if, § 138.
ܐܢܐ I, § 35.
ܐܢܘܢ them *m.*, § 36. 2.
ܐܢܘܫ Enosh.
ܐܢܝܢ them *f.*, § 36. 2.
ܐܢܫ man, one, some one, §§ 90. 4, Rem. 2; 107. 1, 5.

ܐܢܬ thou *m.*
ܐܢܬܝ thou *f.*
ܐܢܬܬܐ woman, § 87, 8.
ܐܣܝܘܬܐ healing.
ܐܣܛܪܛܝܘܛܐ soldier.
ܐܣܟܡܐ figure, form.
ܐܣܦܘܓܐ sponge.
ܐܣܘܪ band.
ܐܣܪ to bind.
ܐܦ also.
ܐܦܠܐ also not, nor.
ܐܦܢ although.
ܐܦܝܢ face, vail, § 87, 9.
ܐܪܒܥܐ four.
ܐܪܒܥܝܢ forty.
ܐܪܡܠܬܐ widow.
ܐܪܥܐ earth.
ܐܪܝܡܠܐ *see* ܪܝܡܠܐ.
ܐܫܕ to pour.
ܐܫܬܡܫ *see* ܫܡܫ.
ܐܫܬܐ foundation.
ܐܬܐ to come.
ܐܬܐ sign, § 86. 3.
ܐܬܘܪ Assyria.
ܐܬܪܐ place.

ܒ.

ܒ in, among.
ܒܐܫ II to be displeased.

ܒܕܪ to scatter.

ܒܗܬ to be ashamed.

ܒܘܠܗ waste.

ܒܘܟܪܐ firstborn, firstling.

ܒܘܪܟܬܐ blessings.

ܒܙܐ to despise.

ܒܙܚ to mock.

ܒܚܕܐ at once.

ܒܛܢ to conceive.

ܒܛܢܐ conception.

ܒܝܐ to console.

ܒܝܡ judgment seat.

ܒܝܢܬ between (before suffixes).

ܒܝܫ evil.

ܒܝܫܘܬܐ evil, wickedness.

ܒܝܬ between.

ܒܝܬܐ house.

ܒܝܬ ܥܢܝܐ Bethany.

ܒܟܐ to weep.

ܒܟܬܐ weeping.

ܒܠܚܘܕ alone.

ܒܢܐ to build.

ܒܣܪܐ flesh.

ܒܣܡ to be pleased, to delight.

ܒܣܡܐ incense.

ܒܣܡܐ ointment.

ܒܣܬܪ behindhand.

ܒܠܥ to swallow.

ܒܥܠ lord.

ܒܥܐ to ask.

ܒܥܘܬܐ request.

ܒܥܝܐ inquirer.

ܒܥܝܪܐ cattle.

ܒܥܠܕܒܒܘܬܐ enmity.

ܒܩܪܐ oxen.

ܒܪܐ son, § 87. 10.

ܒܪܢܐܝܬ filially.

ܒܪܢܫܐ son of man.

ܒܪܐ to create.

ܒܪܘܠܐ beryl.

ܒܪܘܠܚܐ bedellium.

ܒܪܟ to bend.

ܒܪܟ to bless.

ܒܪܡ but.

ܒܪܩܐ lightning.

ܒܬܪ after.

ܒܬܪܟܢ afterwards.

ܓ

ܓܒܐ to choose.

ܓܒܐ side.

ܓܒܠ to form.

ܓܒܪܐ man.

ܓܓܘܠܬܐ Golgotha.

ܓܕܠ to twist.

ܓܕܣܡܢܝ Gethsemene.

ܓܕܦ to blaspheme.

ܓܫܦ to touch.

ܓܘ midst.

ܓܘܕܦܐ blasphemy.

ܓܘܫܡܐ body.

ܓܙܪܐ flock.

ܓܚܟ to laugh.

ܓܝܚܘܢ Gihon.

ܓܝܣܐ thief.

ܓܝܪ for.

ܓܝܪܐ adulterer.

ܓܠܐ to reveal.

ܓܠܙ to defraud.

ܓܠܠܐ wave.

ܓܠܝܠܐ Galilee.

ܓܡܪ to complete.

ܠܓܡܪ entirely.

ܓܢܒ to steal.

ܓܢܣܐ kind.

ܓܥܐ to cry.

ܓܥܬܐ cry.

ܓܥܪ to rebuke.

ܓܦܐ wing.

ܓܦܬܐ vine.

ܓܪܒܐ leper.

ܓܪܡܐ bone.

ܕ.

ܕ that, who, those who. *See* §§ 38, 136, 137. 4. 5.

ܕܒܚ to sacrifice.

ܕܒܚܐ sacrifice.

ܕܒܩ to cleave.

ܕܒܪ to lead.

ܕܒܪܐ field.

ܕܓܠ to lie.

ܕܓܠܘܬܐ lie.

ܕܗܒܐ gold.

ܕܘܒܪܐ regimen, life.

ܕܘܘܢܐ misery.

ܕܘܟܪܢܐ memorial, memory.

ܕܢ to judge.

ܕܘܥܬܐ sweat.

ܕܨ to exult.

ܕܫ to bruise.

ܕܚܠ to fear.

ܕܚܠܬܐ fear.

ܕܚܨ to oppress.

ܕܚܫܐ lictor.

ܕܝܠ own, § 106.

ܕܝܢ but, indeed.

ܕܝܢܐ judgment.

ܕܝܢܐ judge.

ܕܝܢܪܐ denar.

ܕܪ to dwell.

ܕܝܪܐ habitation.

ܕܝܪܝܘܬܐ monastic life.

ܕܝܬܝܩܐ covenant.

ܕܟܐ to purify.

ܕܟܝܐ pure.

ܕܟܪ II to remember.

ܕܟܪܐ male.

ܕܠܚ to disturb.

ܕܠܡܐ is it not? § 132.

ܕܡܐ blood.

ܕܡܘܬܐ likeness.

ܕܡܝܐ price.

ܕܡܝܐ like.

ܕܡܝܟ sleeping.

ܕܡܟ to sleep.

ܕܡܥܬܐ tear.

ܕܡܪ II to wonder.

ܕܢܚ to rise.

ܕܩܠܬ Tigris.

ܕܪܓ grade, ordination.

ܕܪܕܪܐ briers.

ܕܪܟ to come to.

ܕܪܫ to exercise, teach.

ܕܪܬܐ palace.

ܗ.

ܗܐ behold.

ܗܒܝܠ Abel.

ܗܓܡܘܢܐ governor.

ܗܕܐ this, § 37.

ܗܘ that, § 37.

ܗܘܝܘ he it is.

ܗܘ he, § 35.

ܗܘܐ to be, § 127.

ܗܘܝܐ existence.

ܗܝ (ܗܝ) she, § 35

ܗܝ that, § 37.

ܗܢܘܢ them.

ܗܝܟܠܐ temple.

ܗܝܡܢ to believe.

ܗܠܝܢ these, § 37.

ܗܟܘܬ so, likewise.

ܗܡܢ there, therefore.

ܗܟܢܐ thus, so.

ܗܠܟ to go, walk.

ܗܢܐ V to profit.

ܗܢܐ this, § 37.

ܗܢܘܢ they, § 35.

ܗܢܘܢ those, § 37.

ܗܢܝܢ those *f.*, § 37.

ܗܦܟ to return, overturn.

ܗܪܟܐ here.

ܗܫܐ now.

ܘ.

ܘ and, that, when, or.

ܘܝ woe, alas.

ܘܠܐ it is right, necessary.

ܘܥܕ to appoint a time.

ܙ.

ܙܒܕܝ Zebedee.

ܙܒܢ to buy.

ܙܒܢܐ time.
ܙܕܝܩܐ just.
ܙܕܝܩܘܬܐ righteousness.
ܙܗܪ II to take heed.
ܙܗܘܡܐ fetid.
ܙܗܝܐ pure.
ܙܥ to be moved.
ܙܘܥܐ earthquake.
ܙܚܘܪܝܬܐ purple.
ܙܝܬܐ olives.
ܙܟܝܐ pure.
ܙܝܘܐ ornament.
ܙܥܘܪܐ small.
ܙܩܝܦܐ cross.
ܙܩܦ to crucify.
ܙܪܥ to sow.
ܙܪܥܐ seed.

ܚ.

ܚܐܪܐ free, noble.
ܚܒܠ to corrupt.
ܚܒܠܐ corruption.
ܚܒܪܐ neighbor.
ܚܒܫ to include, bind up.
ܚܒܘܫܝܐ life or cell of a recluse.
ܚܓܠܐ chaff.
ܚܓܝܪܐ lame.
ܚܕ one.
ܚܕܘܬܐ joy.
ܚܕܝ to be glad.
ܚܕܝ to make glad.
ܚܕܥܣܪ eleven.
ܚܕܪ to surround.
ܚܕܬܐ new.
ܚܘܐ Eve.
ܚܘܒ to be guilty.
ܚܘܝ to show.
ܚܘܒܐ love.
ܚܘܛܪܐ staff, rod.
ܚܘܝܐ serpent.
ܚܘܝܠܐ Havilah.
ܚܣ to pity.
ܚܪ to look.
ܚܘܪ to make white.
ܚܘܪ white.
ܚܘܪܝܒ Horeb.
ܚܙܐ to see.
ܚܙܘܐ vision, countenance.
ܚܛܐ to sin.
ܚܛܗܐ sin. ܚܛܝܐ sinner.
ܚܛܝܬܐ sin. ܚܛܝܬܢܐ sinful.
ܚܛܘܦܝܐ violence.
ܚܝܐ to live.
ܚܝ living.
ܚܝܐ life.
ܚܝܒ guilty.
ܚܝܘܬܐ animal.
ܚܝܠܐ strength.

ܚܝܠܬܢܐ mighty.
ܚܘܪܐ linen.
ܚܟܡ to know.
ܚܠܐ sweet.
ܚܠܐ vinegar.
ܚܠܝܛ mixed.
ܚܠܡܐ dream.
ܚܠܦ to change.
ܚܠܦ for, instead of.
ܚܡܫܐ five.
ܚܡܬܐ wrath.
ܚܝܠܬܐ anguish.
ܚܢܘܟ Enoch.
ܚܢܢ we.
ܚܢܢܬܐ supplication.
ܚܢܩ to strangle.
ܚܢܩܐ cord.
ܚܣܐ to be innocent.
ܚܣܝܐ holy, sacred.
ܚܣܕ to reproach, revile.
ܚܣܡܐ envy.
ܚܦܛ to urge, incite.
ܚܦܝܛܐܝܬ studiously, carefully.
ܚܩܠܐ field.
ܚܪܒ to dry up, be desolate.
ܚܪܒܐ 1. waste. 2. sword, share.
ܚܪܡ to curse.
ܚܪܫܐ enchanter.
ܚܪܬܐ end.
ܚܫ to suffer.
ܚܫܐ suffering.
ܚܫܒ to impute, reckon.
ܚܘܫܒܐ thought, meditation.
ܚܫܘܟܐ darkness.
ܚܬܐ sister.
ܚܬܡ to seal.

ܛ.

ܛܒܐ report, fame.
ܛܒ very.
ܛܒܬܐ good.
ܛܘܒܐ happiness.
ܛܘܗܡܐ nation, race.
ܛܘܥܝ error.
ܛܘܪ mountain.
ܛܝܒ to prepare.
ܛܝܒܘܬܐ goodness, grace.
ܛܝܡܐ price.
ܛܠܝܐ boy.
ܛܠܝܘܬܐ youth.
ܛܠܠܐ shade.
ܛܠܡ to injure, rob.
ܛܠܢܝܬܐ shadow, demon.
ܛܢܦ to pollute, profane.
ܛܢܦܐ profane.
ܛܢܦܘܬܐ impurity.
ܛܥܐ to err, seduce.
ܛܥܡ to taste, eat.

ܛܪܦ to strike.

ܛܪܦܐ leaf.

ܛܫܐ to hide.

ܝ.

ܝܒܝܫܬܐ dry land.

ܝܒܠ to lead.

ܡܝܒܠܢܐ propagator, successor.

ܝܘܒܠ Jobal.

ܝܒܫ to be dry.

ܝܒܫܐ dry land.

ܐܝܕܐ hand, § 87. 2.

ܝܕܐ to confess, give thanks.

ܝܕܥ to know.

ܝܕܥܬܐ knowledge.

ܝܗܒ to give, § 64. 7.

ܝܗܘܕܐ Judah.

ܝܗܘܕܝܐ Jew.

ܝܘܒܠ Jubal.

ܝܘܠܦܢܐ education.

ܝܘܡܐ day.

ܝܘܡܢܐ daily.

ܝܘܢܢ Jonah.

ܝܘܣܐ Joses.

ܝܘܣܦ Joseph.

ܝܘܦܐ Jopha.

ܝܘܪܩܐ greenness.

ܝܚܝܕ only.

ܝܠܕ to bear.

ܝܠܕܐ child.

ܝܡܐ sea.

ܝܡܐ to swear.

ܝܡܝܢܐ right hand.

ܝܘܡܐ day.

ܝܣܦ to add.

ܝܥܐ to spring up.

ܝܥܩܘܒ Jacob.

ܝܥܪܐ forest, thorn.

ܝܩܕ to burn.

ܝܩܝܪܐ precious.

ܝܩܪ to honor.

ܐܝܩܪܐ honor.

ܝܪܒ to be great.

ܝܪܘܬܐ heir.

ܝܪܬܘܬܐ inheritance.

ܝܫܛ to extend.

ܝܫܘܥ Jesus.

ܝܬ Gen. 1. 1, § 89*c*.

ܝܬܒ to sit.

ܝܬܝܪ more.

ܝܬܝܪܐܝܬ more.

ܝܬܡܐ orphan.

ܝܬܪ to profit.

ܟ.

ܟܐܒܐ sorrow.

ܟܐܦܐ stone.

ܟܒܪ but.

ܟܒܰܫ to subdue.

ܟܶܒܪܺܝܬܳܐ sulphur.

ܟܰܕ when, while.

ܟܕܽܘ it is sufficient.

ܟܳܗܢܳܐ priest.

ܟܰܘܳܐ window.

ܟܽܘܒܶܐ thorns.

ܟܰܘܟܒܳܐ star.

ܟܽܘܪܣܝܳܐ throne.

ܟܽܘܬܺܝܢܳܬܳܐ tunics.

ܟܽܘܫ Cush.

ܟܝܳܢܳܐ nature.

ܟܺܝܬ indeed.

ܟܽܠ all, § 108.

ܟܠܳܐ to withhold, restrain.

ܟܠܺܝܠܳܐ crown.

ܟܠܰܡܺܝܣ Chlamys.

ܟܡܳܐ how.

ܟܡܰܪ to be sad.

ܟܢܽܘܫܬܳܐ congregation.

ܟܶܢܳܪܳܐ cithara.

ܟܢܰܫ to assemble.

ܟܶܢܫܳܐ assembly, collection.

ܟܣܳܐ to cover.

ܟܳܣܳܐ cup.

ܟܶܣܦܳܐ silver.

ܟܦܰܪ to deny.

ܟܪܳܐ to be sad.

ܟܪܽܘܒܳܐ Cherub.

ܟܳܪܽܘܙܽܘܬܳܐ preaching.

ܟܪܰܙ to preach.

ܟܰܪܝܳܐ sad.

ܟܪܺܝܗܳܐ sick, infirm.

ܟܰܪܝܽܘܬܳܐ sadness.

ܟܪܶܟ to surround, lead around.

ܟܰܪܡܳܐ vineyard.

ܟܰܪܣܳܐ belly.

ܟܫܰܠ to stumble.

ܟܫܰܦ to beseech.

ܟܬܰܒ to write.

ܟܬܳܒܳܐ book.

ܟܶܬܳܢܳܐ linen.

ܟܬܰܫ to strive.

ܠ.

ܠ to § 123 sq.

ܠܳܐ not.

ܠܐܳܐ to labor, be weary.

ܠܶܒܳܐ heart.

ܠܒܶܫ to clothe.

ܠܒܽܘܫ clothing.

ܠܰܓܬܳܐ dish.

ܠܶܓܝܽܘܢܶܝܢ legions.

ܠܶܘܺܝ Levi.

ܠܛܰܐ to curse.

ܠܽܘܩܕܰܡ before.

ܠܘܳܬ to, with.

ܠܘܳܛܳܬܳܐ curses.

ܠܰܚܡܳܐ bread.

ܠܰܥܓܰܠ quickly, immediately.

ܠܰܝܬ there is not, § 65, 128.

ܠܺܠܝܳܐ night.

ܠܰܡ indeed, forsooth.

ܠܰܡܶܟ Lamech.

ܠܶܣܛܳܝܳܐ thief.

ܠܥܶܠ above.

ܠܥܰܣ to eat.

ܠܶܫܳܢܳܐ tongue.

ܡ.

ܡܳܐ what?

ܡܶܐܟ݂ܽܘܠܬܳܐ food.

ܡܳܐܢܳܐ vessel, vestment.

ܡܰܒܽܘܥܳܐ fountain.

ܡܰܓܕܠܳܝܬܳܐ Magdalene.

ܡܰܓܳܢ in vain.

ܡܰܕܒܚܳܐ altar.

ܡܰܕܒܪܳܐ wilderness.

ܡܕܺܝܢܬܳܐ city.

ܡܶܕܶܡ anything § 109.

ܡܰܕܢܚܳܐ east.

ܡܰܕܢܚܳܐ orient, east.

ܡܰܕܢܚܳܝܳܐ eastern.

ܡܰܘܠܳܕܳܐ birth.

ܡܰܘܡܳܬ݂ܳܐ oaths.

ܡܽܘܢ what?

ܡܽܘܫܶܐ Moses.

ܡܰܘܬܳܐ death.

ܡܚܳܐ to smite.

ܡܶܚܕܳܐ straitway.

ܡܚܽܘܝܳܐܝܶܠ Mehujael.

ܡܚܳܪ morrow.

ܡܰܚܫܽܘܠܳܐ wave, billow.

ܡܛܳܐ to come.

ܡܶܛܽܠ on account of.

ܡܶܛܽܠ ܕ because that.

ܡܛܰܠܬܳܐ booth, shade.

ܡܛܰܥܝܳܢܳܐ deceiver.

ܡܶܛܪܳܐ rain.

ܡܰܝܳܐ water.

ܡܝܰܠܕܳܢܳܐ propagator, successor.

ܡܺܝܛܪܽܘܦܳܘܠܺܝܛܳܐ Metropolitan.

ܡܰܝܩ to deride.

ܡܺܝܬ to die.

ܡܺܝܬܳܐ dead.

ܡܝܰܬܪܳܬܳܐ meliora, virtues.

ܡܶܟܳܐ yet, hence.

ܡܰܟܺܝܟܳܐ humble.

ܡܰܟܺܝܟܳܐܝܬ humbly.

ܡܰܟܺܝܟܽܘܬܳܐ humility.

ܡܶܟܺܝܠ now.

ܡܟܰܪ to betroth.

ܡܠܳܐ to be full.

ܡܶܠܬܳܐ word.

ܡܰܠܰܐܟ݂ܳܐ messenger, angel.

ܡܰܠܰܐܟ݂ܺܝ Malachi.

ܡܠܚܐ sailor.

ܡܠܟ to counsel.

ܡܠܟܐ king.

ܡܠܟܐ counsel.

ܡܠܟܘܬܐ kingdom.

ܡܠܬܐ word.

ܡܠܠ to speak.

ܡܡܠܠܐ word.

ܡܠܦܢܐ teacher.

ܡܡܬܘܡ ever, at all.

ܡܢ from, more than.

ܡܢ who? ܡܢܘ who is?

ܡܢܐ what? § 39, 132.

ܡܢܘ what is?

ܡܢܐ to take part, be numbered.

ܡܛܐ to come, bring.

ܡܣܟܢ to become poor.

ܡܣܟܢܐ poor.

ܡܣܪܗܒܐ quick, bold.

ܡܥܕܪܢܐ helper.

ܡܥܝܢܐ spring, fountain.

ܡܥܣܪܐ tithes.

ܡܥܪܒܐ setting.

ܡܥܪܬܐ cave.

ܡܦܝܣܢܐ supplicator.

ܡܨܥܬܐ midst.

ܡܩܒܪ burial.

ܡܪܕܐ way.

ܡܪܘܝܢܐ intoxicating.

ܡܪܚܡܢܐ merciful.

ܡܪܚܡܢܘܬܐ mercy.

ܡܪܚܦܢܐ clement.

ܡܪܛܘܛܐ garment.

ܡܪܝܐ Lord.

ܡܪܝܡ Mary.

ܡܪܝܪܐܝܬ bitterly.

ܡܪܪܬܐ gall.

ܡܪܬܝܢܘܬܐ admonition.

ܡܫܘܚܬܐ stature, age.

ܡܫܚܐ oil.

ܡܫܝܚܐ Messiah.

ܡܫܝܚܝܐ Messianic.

ܡܫܟܚ able, possible.

ܡܫܟܐ skin.

ܡܫܟܢܐ tent, house.

ܡܫܠܡܢܐ traitor.

ܡܫܡܠܝܢܐ perfect, whole.

ܡܫܬܝܐ drink.

ܡܬܘܫܐܝܠ Methusael.

ܢ.

ܢܐ now.

ܢܒܐ to prophecy.

ܢܒܝܐ prophet.

ܢܣܩ to kindle.

ܢܓܕ to smite, beat.

ܢܗܪ to shine.

ܢܓܝܪܐ long.

ܢܓܰܪ to be long.
ܢܕܰܪ to vow.
ܢܶܕܪܳܐ vow.
ܢܗܰܪ to shine.
ܢܰܗܪܳܐ river.
ܢܰܗܺܝܪܳܐ light.
ܢܽܘܕ Nod.
ܢܳܕ to move, wander.
ܢܽܘܗܪܳܐ light.
ܢܳܚ to rest.
ܢܽܘܟܪܳܝܳܐ stranger.
ܢܽܘܢܳܐ fish.
ܢܽܘܪܳܐ fire.
ܢܚܳܫܳܐ brass.
ܢܚܶܬ to go down.
ܢܰܚܬܳܐ garment.
ܢܛܽܘܪܽܘܬܳܐ injunction.
ܢܛܰܦ to distil, instil.
ܢܛܰܪ to watch, observe.
ܢܺܝܚܳܐܝܺܬ quietly.
ܢܺܝܢܘܶܐ Nineveh.
ܢܺܝܪܳܐ yoke.
ܢܺܝܫܳܐ sign, purpose.
ܢܶܟܠܳܐ guile.
ܢܰܟܦܽܘܬܳܐ chastity.
ܢܳܡܽܘܣܳܐ law.
ܢܳܡܽܘܣܳܐܝܺܬ lawfully.
ܢܣܳܐ to try, tempt.
ܢܣܰܒ to take, receive.

ܢܶܣܝܽܘܢܳܐ temptation.
ܢܣܰܟ to pour out.
ܢܰܥܡܳܐ Naamah.
ܢܦܰܫ to breath.
ܢܦܰܠ to fall.
ܢܦܰܩ to go out.
ܢܦܰܨ to break.
ܢܰܦܫ soul.
ܢܨܰܒ to plant.
ܢܶܨܒܬܳܐ plant.
ܢܳܨܪܳܝܳܐ Nazarene.
ܢܶܩܒܬܳܐ female.
ܢܰܩܕܳܐ pure.
ܢܩܺܝܪ hewn.
ܢܩܶܦ to cleave to.
ܢܩܰܫ to knock.
ܢܫܰܒ to breathe.
ܢܶܫܒܬܳܐ breath.
ܢܫܰܩ to kiss.

ܣ.

ܣܺܐܡܳܐ silver.
ܣܒܰܪ to think.
ܣܰܒܪܳܐ hope.
ܣܒܰܪܬܳܐ gospel.
ܣܰܓܺܝ much, great.
ܣܰܓܺܝܐܳܐ much, many.
ܣܓܳܐ to multiply.
ܣܓܶܕ to worship.

ܣܗܕ to witness.
ܣܘܓܐܐ multitude.
ܣܘܥܕܐ colloquy, word.
ܣܘܟܪܐ bar.
ܣܡ to put.
ܣܛܐ to incline, sin.
ܣܝܒܪ to endure.
ܣܝܦܐ sword.
ܣܟܐ to expect.
ܣܟܠ to be wise.
ܣܟܠܘܬܐ trespass.
ܣܟܪܝܘܛܐ Iscariot.
ܣܠܐ to reject.
ܣܠܩ to go up.
ܣܡܝܟ lying.
ܣܡܠܐ left, left hand.
ܣܢܐ to hate.
ܣܥܪ to do, happen.
ܣܦܐ threshold.
ܣܦܬܐ lip.
ܣܦܣܪܐ sword.
ܣܦܪ to receive the tonsure.
ܣܦܪܐ book.
ܣܦܪܐ scribe.
ܣܩܐ sackcloth.
ܣܪܝܩܘܬܐ vanity.
ܣܪܝܩܐ vain.
ܣܪܝܩܐܝܬ in vain.
ܣܬܪ to destroy.

ܥ.

ܥܐܕܐ feast.
ܥܒܕ to make.
ܥܒܕܐ workman, slave.
ܥܒܕܐ work.
ܥܒܪ to pass over, transgress.
ܥܒܪܐ the uttermost part.
ܥܒܪܝܐ Hebrew.
ܥܓܠ to roll.
ܥܓܠ quickly.
ܥܓܠܐ calf.
ܥܕ until.
ܥܕܐ Ada.
ܥܕܟܝܠ as yet.
ܥܕܠܐ before that.
ܥܕܡܐ until.
ܥܕܢ Eden.
ܥܐܕܐ feast.
ܥܕܬܐ church.
ܥܕܬܢܝܐ ecclesiastical.
ܥܗܝܕܐ memorable.
ܥܝܕ to be customary.
ܥܘܡܩܢܝܐ grievous.
ܥܘܪܕܢܐ thorn.
ܥܘܝܪܐ blind.
ܥܘܠܐ iniquity.
ܥܘܠܐ evil-doer.
ܥܘܡܩܐ depth.
ܥܘܦܐ branch.

ܥܩ to be sad, anxious.

ܥܪ to watch.

ܥܛܐ to wipe out.

ܥܝܪܕ Edar (Irad).

ܥܝܢܐ eye.

ܥܕܝܢ before.

ܥܠ upon, over, against, at, unto, for, on account of.

ܥܠ ܕ because.

ܥܠ to go in.

ܥܠܡ ever, age, world.

ܥܠܥܝܢ above.

ܥܠܡܢܝܐ worldly, secular.

ܥܠܬܐ cause, accusation.

ܥܡ with.

ܥܡܐ people.

ܥܡܕ to baptize.

ܥܡܘܕܪܐ farmer.

ܥܡܠܢܐ laborious.

ܥܡܠ to toil.

ܥܡܠܐ toil.

ܥܡܪ to dwell.

ܥܢܐ to answer.

ܥܢܐ flock.

ܥܢܢܐ cloud.

ܥܣܒܐ herb.

ܥܝܣܘ Esau.

ܥܣܪܝܢ twenty.

ܥܦܪܐ dust.

ܥܩܐ grief, anxiety.

ܥܩܒܐ heel.

ܥܩܪܐ root.

ܥܪܒܐ sheep.

ܥܪܘܒܬܐ evening.

ܥܪܝܡ cunning, subtle.

ܥܪܛܠܝ naked.

ܥܪܩ to flee.

ܥܫܢ to be strong.

ܥܬܝܕܐ future.

ܥܬܝܪܐ rich.

ܦ.

ܦܐܪܐ fruit.

ܦܓܥ to meet.

ܦܓܪܐ body.

ܦܓܪܢܐܝܬ corporally.

ܦܓܪܢܝܐ corporal.

ܦܕܢܐ plough.

ܦܘܡܐ mouth.

ܦܣ to persuade.

ܦܘܪܥܢܐ tribute, remuneration.

ܦܘܩܕܢܐ commandment.

ܦܘܪܫܢܐ separation, judgment.

ܦܫ to cease, remain.

ܦܘܚܡܐ comparison.

ܦܚܪܐ potter.

ܦܛܝܪܐ unleavened bread.

ܦܝܠܛܘܣ Pilate.

ܦܝܣܐ supplication.
ܦܝܫܘܢ Pishon.
ܦܠܓ to doubt, divide.
ܦܠܚ to till, work.
ܦܠܛ to cast out.
ܦܠܢ a certain one.
ܦܠܓܐ occasion, opportunity.
ܦܢܐ to turn.
ܦܢܝܐ turning.
ܦܣܬܐ sole.
ܦܣܐ lot.
ܦܣܩ to cut off, break.
ܦܨܐ to free, liberate.
ܦܨܚܐ passover.
ܦܩܕ to command.
ܦܩܚ useful, tolerable.
ܦܩܥܬܐ plain.
ܦܪܐ to be fruitful.
ܦܪܓܠܐ whip.
ܦܪܕܝܣܐ Paradise.
ܦܪܙܘܡܐ apron.
ܦܪܙܠܐ iron.
ܦܪܚ to fly.
ܦܪܚܬܐ bird.
ܦܪܛܘܪܝܢ pretorium.
ܦܪܥ to avenge.
ܦܪܨܘܦܐ face.
ܦܪܩ to go away, free.
ܦܪܫ to separate, assign.
ܦܪܬ Euphrates.
ܦܪܬܐ dung.
ܦܫܩ to expound.
ܦܬܓܡܐ word.
ܦܬܘܪܐ table.
ܦܬܚ to open.

ܨ.

ܨܒܐ to wish, will.
ܨܒܝܢܐ will.
ܨܒܝܢܐܝܬ willingly.
ܨܒܥ to dip.
ܨܗܝܘܢ Zion.
ܨܘܐܐ filth.
ܨܘܠܦܐ wound.
ܨܘܡܐ fasting.
ܨܬ to hearken.
ܨܚ to burn, be hot.
ܨܝܕ unto, by.
ܨܠܐ Zillah.
ܨܠܝ to pray.
ܨܠܘܬܐ prayer.
ܨܠܡܐ image.
ܨܚܢܘܬܐ foulness.
ܨܦܪܐ dust.
ܨܦܬܐ care.
ܨܪܐ to tear.
ܨܪܦ to refine.

ܩ.

ܩܐܝܢ Cain.

ܩܒܘܪܐ sepulchre.

ܩܒܠ to receive.

ܩܢܐ to possess.

ܩܕܝܡ east.

ܩܕܝܫܐ holy.

ܩܕܡ to go before, anticipate.

ܩܕܡ before.

ܩܕܡܝܐ first.

ܩܕܡܝܬ first.

ܩܕܫ to sanctify.

ܩܘܐ to wait, remain.

ܩܘܢܝ possessor.

ܩܘܒܠ before.

ܩܘܕܫܐ holiness.

ܩܘܡ to rise.

ܩܘܪܒܢܐ offering.

ܩܘܪܝܢܝܐ Cyrenian.

ܩܘܫܬܐ truth.

ܩܛܘܠܐ killer.

ܩܛܠ to kill.

ܩܛܥ to cut off.

ܩܛܡܐ ashes.

ܩܝܛܘܢܐ chamber, room.

ܩܝܡܬܐ resurrection.

ܩܝܦܐ Caiaphas.

ܩܝܬܪܐ cithara.

ܩܠܐ voice.

ܩܠ to be light.

ܩܠܝܠ little.

ܩܠܝܬܐ cell.

ܩܠܣ to praise, celebrate.

ܩܠܪܝܩܝܐ clerical.

ܩܢܐ to acquire, possess.

ܩܢܘܒܝܐ cenobite.

ܩܢܝܐ reed, cane.

ܩܢܝܢܐ possession, gift.

ܩܢܛܪܘܢܐ Centurion.

ܩܣܛܪܛܝܘܛܐ soldier.

ܩܥܐ to cry.

ܩܦܚ to buffet.

ܩܨܐ to break.

ܩܪܐ to call.

ܩܪܐܐ cucumber.

ܩܪܒ to be near.

ܩܪܝܬܐ city.

ܩܨܝܐ piece.

ܩܪܩܦܬܐ skull.

ܩܫܝܫܐ elder.

ܪ.

ܪܒܐ to be great, to multiply.

ܪܒܐ great.

ܪܒܝ Rabbi.

ܪܒܘܬ myriads.

ܪܒܝܥ crouched, laid.

ܪܓܙ to be angry.

ܪܓܝܓܐ desired, desirable.
ܪܓܫ to be tumultuous.
ܪܓܫ to perceive.
ܪܓܬܐ desire.
ܪܕܐ to go, instruct.
ܪܗܛ to run.
ܪܗܛܐ course.
ܪܘܒܐ strife, tumult.
ܪܘܓܙܐ wrath.
ܪܘܚ to refresh.
ܪܘܚܐ wind, spirit.
ܪܘܚܩܐ afar.
ܪܘܪܒܐ many, § 86. 1.
ܪܘܪܒܢܐ magnates, § 86. 1.
ܪܚܡ to love, have mercy.
ܪܚܡܐ mercy.
ܪܚܦ to brood.
ܪܚܩ to be far.
ܪܚܫ to creep, to move oneself.
ܪܚܫܐ creeping things.
ܪܝܫܐ head.
ܪܡܐ to cast.
ܪܡܐ high.
ܪܡܬܐ Aramathea.
ܪܡܫܐ evening.
ܪܢܐ to meditate.
ܪܥܐ to feed, think.
ܪܥܝܐ shepherd.
ܪܩ to spit.
ܪܩܝܥܐ firmament.
ܪܬܝܬܐ trembling.

ܫ

ܫܐܠ to ask.
ܫܐܠܬܐ request.
ܫܒܐ week.
ܫܒܒܐ neighbor.
ܫܒܚ to praise.
ܫܒܛܐ rod.
ܫܒܝܥܝܐ seventh.
ܫܒܥ seven.
ܫܒܥܐ seven.
ܫܒܥܝܢ seventy.
ܫܒܩ to leave.
ܫܒܬܐ week, sabbath.
ܫܓܘܫܝܐ sedition.
ܫܕܐ to cast, throw away.
ܫܕܪ to send.
ܫܗܪ to watch.
ܫܘܐ to be worthy.
ܫܘܒܐ heat.
ܫܘܒܩܢܐ remission.
ܫܘܓ to wash.
ܫܘܕܥܢܐ sign, inducement.
ܫܘܠܛܢ power, ruling.
ܫܘܠܡܐ end.
ܫܘܩܦܐ blow.
ܫܘܩܪܐ lie.

ܫܰܘܬܶܦ to espouse, marry.
ܫܰܘܬܳܦܳܐ companion.
ܫܝܺܛܳܐ vexed.
ܫܰܚܶܪ to compel.
ܫܝܺܛ cursed.
ܫܺܝܫܬܳܐ alabaster box.
ܫܝܘܿܠ Sheol.
ܫܝܬ Seth.
ܫܟܰܚ to find, be able, possible.
ܫܟܝܺܒ asleep.
ܫܠܳܐ to rest, be calm.
ܫܶܠܝܳܐ rest, sleep.
ܫܠܰܚ to send, to take off.
ܫܠܶܛ to rule.
ܫܰܠܝܺܛ lawful, ruler.
ܫܰܠܝܺܛܳܢܳܐ ruler.
ܫܠܳܡ peace.
ܫܠܶܡ to finish, Ap. to betray.
ܫܶܡ name.
ܫܡܰܛ to draw.
ܫܡܰܝܳܐ heaven.
ܫܰܡܺܝܢܳܐ fatling.
ܫܡܰܝܳܢܳܐ heavenly.
ܫܡܰܥ to hear.
ܫܶܡܥܽܘܢ Simon.
ܫܰܡܶܫ to minister.
ܫܶܡܫܳܐ sun.
ܫܢܳܐ to depart.
ܫܢܰܝܳܐ years.
ܫܢܺܝܢ years.
ܫܢܳܢܳܐ point.
ܫܥܳܐ to narrate.
ܫܳܥܬܳܐ hour.
ܫܰܘܺܝ to make level.
ܫܰܦܺܝܪ beautiful, good.
ܫܦܰܟ to pour.
ܫܦܰܪ to be good.
ܫܰܦܪܳܐ morning.
ܫܩܳܐ to irrigate.
ܫܩܰܠ to take away.
ܫܪܳܐ to cast away, begin.
ܫܰܪܒܳܐ history.
ܫܰܪܒܬܳܐ family.
ܫܒܽܘܫܬܳܐ branch, vine.
ܫܰܪܺܝܪܳܐ true.
ܫܰܪܺܝܪܳܐܺܝܬ truly.
ܫܰܪܟܳܐ rest.
ܫܶܬ six.
ܫܬܳܐ to drink.
ܫܰܬܺܝܩ silent.
ܫܬܺܝܬܳܝܳܐ sixth.

ܬ.

ܬܺܐܢܬܳܐ fig tree.
ܬܒܺܝܪ broken.
ܬܒܰܪ to break.
ܬܳܓܳܐ crown.
ܬܶܟܣܳܐ order.

ܬܕܐܐ grass.

ܬܗܘܡܐ abyss.

ܬܐܒ to repent.

ܬܒ to turn.

ܬܘܒ again.

ܬܘܒܠܩܝܢ Jubal Cain.

ܬܘܕܝܬܐ thanksgiving.

ܬܘܗ formless.

ܬܘܟܠܢܐ trust, confidence.

ܬܘܠܕܐ generation.

ܬܘܠܥܐ worm.

ܬܘܪܐ cattle.

ܬܚܘܡܐ boundary.

ܬܚܝܬ under.

ܬܚܘܬ under.

ܬܚܬ under.

ܬܟܝܠܐ trusting.

ܬܟܝܠܐܝܬ trustingly.

ܬܟܫܦܬܐ prayer.

ܬܠܓܐ snow.

ܬܠܡܕ to teach.

ܬܠܡܝܕܐ disciple.

ܬܠܬ three.

ܬܠܬܝܢ thirty.

ܬܡܢ there.

ܬܢܘܪܐ furnace.

ܬܢܚܬܐ sigh.

ܬܢܝܢܐ second.

ܬܢܝܢܐ dragon.

ܬܩܢ to prepare.

ܬܪܝܢ two.

ܬܪܒܝܬܐ growth, increase.

ܬܪܝܨܐ right.

ܬܪܝܨܘܬܐ uprightness.

ܬܪܢܓܠܐ cock.

ܬܪܥܐ door.

ܬܪܥܣܪ twelve.

ܬܪܫܝܫ Tarshish.

ܬܪܬܝܢ two.

ܬܪܬܥܣܪܐ twelve.

ܬܫܥ nine.

ܬܬܬ *see* [illegible].

MANUAL.

PART I.

LESSON ONE. Gen. I. 1.

1. Notes.

1. ܒܪܺܝܫܺܝܬ݂—*b*[e]*rî-shith* (two syllables).—In beginning.

(1) *Six letters:*—ܒ (*b*); ܪ (*r*); ܝ (*y*), occuring twice, both times silent after ̈; ܫ (*sh*); ܬ݂ (*th*, as in *thin*).

(2) *Three vowel sounds:*—([e]) a half-vowel, like *e* in *below* or the obscure vowel of Webster's Dictionary. There is no sign for this half-vowel which corresponds to vocal Sh[e]wa in Hebrew, see § 9; ܫܺ (î) like *i* in *machine*; ܫܻ (î), the same as the last, since ̈ may be written either above or below the letter to which it belongs. § 6. 4.

Note.—ܪܻܫܻܝܬ݂ is the Nestorian form.

2. ܒܪܳܐ—*b*[e]*ro'* (one syllable), (*he created*).

(1) *Three letters:*—ܒ (*b*); ܪ (*r*); ܐ ('), called Olaph, not pronounced but quiescing in the preceding vowel. § 2 (1) *b*.

(2) *Two vowel sounds:*—([e]), see 1 (2); ̊ (*o*) like *o* in *note*.

(3) Note that the half-vowel does not make a syllable, but every full vowel does § 16. 1.

3. ܐܰܠܳܗܳܐ—*'a-lo-ho'* (three syllables), *God*.

(1) *Four letters:*—ܐ ('); ܠ (*l*); ܗ (*h*); ܐ ('); see 2 (1).

(2) *Three vowel sounds:*—̰ (*a*) like *a* in *at*; ̊ (*o*) occuring twice, see 2 (2).

4. ܝܳܬ݂—*yoth*,—not translated, but sign of direct object § 89 *c*.

5. ܫܡܰܝܳܐ—*sh*[e]*ma-yo'* (two syllables)—*the heavens*.

(1) *Four letters:*—ܫ (*sh*); ܡ (*m*); ܝ (*y*); ܐ (').

(2) *Three vowel sounds:*—(ᵉ) see 1 (2); ܰ (*a*) see 3 (2); ܳ (*o*) see 2 (2).

6. ܘܝܳܬ—*wᵉyoth* (one syllable), *and* followed by the sign of the direct object, see 4.

One new letter ܘ (*w*), like *w* in *water*.

7. ܐܰܪܥܳܐ—*'ar-'ol* (two syllables), *the earth.*

Four letters: two Olaphs, see 2 (1); ܪ (*r*), see ܪ 1 (1). (The form ܪ is used at the beginning of a word, or after a letter which does not connect with following letters; the same difference of form as to the Olaph: § 4. 4); ܥ ('), not pronounceable, called Ê, § 3.

2. Observations.

1. The letters in this verse are (1) ܐ, (2) ܒ, (3) ܗ, (4) ܘ, (5) ܝ, (6) ܬ, (7) ܡ, (8) ܬ, (9) ܪ, ܪ, (10) ܫ, (11) ܥ.

2. The vowel signs are (1) ܰ, (2) ܳ, (3) ܶ, all of which may be written either above or below the line. § 6. 4.

3. The vowel sounds are (1) ᵉ, (2) *ă*, (3) *o*, (4) *î*.

4. ܝܳܬ, Targum יָת, is found in a dozen passages of the Old Testament in the Peshitto version. § 89 *c*.

5. ܒ and ܘ are never written separately, being always prefixed to the following word. § 34.

6. The definite state is denoted by the ending ܳܐ, which corresponds to the Hebrew article. § 76.

7. Every syllable begins with a consonant. § 15. 2.

8. Notice that all of the consonants have their direct equivalents in Hebrew, except ܥ which here stands for Hebrew Tsodhe.

3. Grammar Lesson.

(1) §§ 1—4, 9—11, 34. 1. inseparable particles

(2) Gender, number and state of nouns. § 76.

4. Word Lesson.

ܓܒܰܠ *he formed.* ܥܒܰܕ *he made.*

ܐܶܡܰܪ *he said.* ܘ *and.*

ܒ *in.*

ܟܬܒ *he wrote.*

ܪܫܝܬ *beginning.*

ܝܬ sign of the definite object.

5. EXERCISES.

1. And beginning. 2. And he formed the heavens. 3. He made the earth and the heavens. 4. God is in the heavens. 5. In the beginning God said. 6. He wrote the beginning.

7. Write out the Syriac of Lesson One in Hebrew characters and note the differences of the languages.

8. Translate the first lesson from Hebrew into Syriac.

9. Retranslate literally into Hebrew. (*Note.*—In these last two exercises, English may be substituted for Hebrew. They may better be written on the board.)

LESSON TWO. Gen. I. 2.

1. NOTES.

8. ܘܐܪܥܐ—*w'ar-'ā* (two syllables), *and the earth.* The vowel ̄ is thrown back on the unvowelled ܘ and Olaph quiesces in the vowel, §§ 25. 1. (2), 34. 2.

9. ܗܘܬ—*hewath* (one syllable), (*she*) *was.* The ܬ is the sign of the feminine; the masculine is ܗܘܐ.

10. ܬܘܗ—*tuh, a desolation.*

(1) ܬ with the dot above is *t*; with the dot below as in ܝܬ is *th*, § 10.

(2) The vowel ܘ is *u* pronounced like *oo* in *fool*, § 6. 3. (3). ܘ is a vowel letter, § 5. 2 & § 6. 5.

(3) For the point above ܘ, see § 6.

No further attention need be paid to the points above and below the ܘ.

11. ܘܒܘܗ—*webhuh* (one syllable), *and a waste.*

(1) ܒ is not *b* (ܒ) but *bh*=*v* in *vote*, § 10. 1. (2).

(2) ܗ is always a consonant in Syriac and never a vowel letter, § 25. 4.

12. ܘܚܫܘܟܐ—*weḥesh-shu-khā* (three syllables) *and (the) darkness.*

(1) ܘ (*w*); ܚ (*ḥ*=ח) like *ch* in *loch*; ܫ (*sh*) is here doubled because it is of a nominal form which doubles the 2d radical, § 72. 2. (6); ܘ=*u*; ܟ=*kh* (כ); ܐ (')=א.

(2) Although this noun has the ending ܐ̊ (see observation 6), it is indefinite, § 93. 2.

13. ܥܰܠ ܐܰܦܰܝ̈—*'al-ap-p'y, upon the faces of.*

(1) Notice *l* final=ܠ while *l* initial or medial=ܠ, § 4. 1.

(2) ܦ̇ after a consonant=*p*; but after a vowel=*pp*; ܦ̣ after a vowel or half-vowel=*ph* or *f*, § 10.

(3) *ay* forms a diphthong and denotes the construct plural, § 8. 2. (1). § 76. 3.

(4) The two dots over ܝ are the sign of the plural, § 13. 1.

14. ܬܗܽܘܡܳܐ—*t'hu-mo* (two syllables), *the abyss* (תְּהוֹם).

(1) The first syllable begins with two consonants, but between them is a half vowel, § 16. 2.

(2) Both syllables are open, § 17. 1.

(3) ܘ quiesces in ܽ, § 6. 5, § 25. 2.

15. ܘܪܽܘܚܶܗ—*w'ru-ḥeh* (two syllables), *and his spirit* (and the spirit of him).

(1) Five consonants, one vowel letter § 5. 2, two vowel signs § 6.

(2) The form consists of the conjunction ܘ, the noun ܪܽܘܚ, and the pron. suffix 3rd sing. masc. ܶܗ, §§ 34, 36.

(3) The vowel ܽ is always written above the consonant, the others may be written below, § 6. 4.

(4) The suffix ܶܗ is used for emphasizing the first of two definite nouns, the second being generally preceded by ܕ, § 97. B.

(5) ܕܰܐܠܳܗܳܐ—*da'-lo-ho who (is) God*, is in apposition with the suffix in ܪܽܘܚܶܗ, § 97. B.

(6) ܕ is the relative pronoun, § 38.

(7) The Olaph after ܕ throws back its vowel and quiesces, §§ 32. 2, 25. 1. (2).

(8) The final ܐ̊ is the sign of the emphatic state of the masc. singular, § 76. 1.

(9) For the form of the noun, see § 69. 2.

16. ܡܪܰܚܦܳܐ—*m'raḥ-ḥ'pho* (two syllables), *brooding.*

(1) ܦ̣=*ph*, ܦ̇=*p* or *pp*.

(2) ܡ prefixed denotes the participle, § 74.

(3) ܐ is the sign of the feminine singular in the absolute state, § 76. 2.

(4) ܚ is doubled, this being in the intensive stem, called Paʿel, § 41. 2.

17. ܡ̈ܝܐ—*ma-yoh the waters.*

(1) The two dots are called Rebbuy and denote the plural, § 13. 1.

(2) ܐ denotes the emphatic or definite state, which takes the place of the article in Hebrew, §§ 86. 16, 87. 22.

18. ܘܶܐܡܰܪ—*weʾ-mar, and he said.*

(1) There is no Waw conversive in Syriac.

(2) ܘܶܐܡܰܪ is composed of Waw and ܐܶܡܰܪ, the Olaph throwing back its vowel and quiescing, see 8 above.

(3) ܐܶܡܰܪ is the 3rd masc. sing. of the Perfect of the simple, or Pᵉal, stem.

(4) Initial Olaph always takes a helping vowel, § 55. 1, *Rem.* 1.

19. ܢܶܗܘܶܐ—*neh-wê, let there be.*

(1) The ܢ (Nun) indicates the Imperfect 3rd person, § 45. 1, *Rem.* 2—4.

(2) The root is ܗܘܐ § 60. 3. Comp. ܗܘܳܬ (9) *she was.*

(3) In the 3rd person, the Imperfect is employed as a Jussive § 114. 1.

20. ܢܘܗܪܐ—*nuh-roh light.*

(1) ܘ is a vowel letter, as in 10 above.

(2) As to form, this noun is in the emphatic state and should be definite; but as to fact, the emphatic state is often used when the noun is indefinite, § 93. 2.

(3) The noun is a *u* class segholate, § 67. 1 *c*.

21. ܘܰܗܘܳܐ—*wa-hᵉwoh and there was.*

(1) There is no Waw conversive.

(2) This is the 3rd masc. sing. Perfect Pᵉal. Comp. (19 (2)) and (9).

(3) Waw receives the helping-vowel *a*, and forms with it a half-open syllable, §§ 16. 4, 32. 2, 33. 3.

2. Observations.

9. The new letters in this verse are: (1) ܚ, (2) ܝ (ܝ), (3) ܡ, (4) ܪ, (5) ܢ.

10. The new forms of letters occurring are: (1) ܠ (ܠ), (2) ܠ (ܠ).

11. The new vowels and diphthongs are: (1) ܘܳ, (2) ̈, (3) ܶ (4) ܐ.

12. The conjunction Waw may be written (1) without a vowel sign,

having merely the half-vowel ^{e} between it and the next letter, or (2) with a helping *a* as in ܐܘܣܘܿ, or (3) when it is followed by a word beginning with Olaph, it draws the vowel to itself the Olaph quiescing, § 34.

13. Syllables ending in a vowel sound are called *open;* ending in a consonant, they are called *closed;* ending in a vowel followed by a consonant with a half-vowel, they are called half-open, § 16.

14. Observe the difference between ܒ݁ (*b*) and ܒ݂ (*bh*); ܚ (*ḥ*) and ܗ (*h*); ܘ (*w*) and ܘܼ (*u*); ܦ (*ph*) and ܦ݁ (*p*); ܬ݁ (*t*) and ܬ݂ (*th*); ܕ (*d*) and ܪ (*r*); ܥ (ʿ) and ܠ (*l*); ܟ (*kh*) and ܒ (*bh*). § 4. 3.

15. Observe that ܝ quiesces in ‍ ; ܘ in ‍ ; ܐ in ‍ , ‍ and ‍ . § 6. 5. *Rem.* and § 24.

16. The Syriac verbal inflection distinguishes number and tense.

17. The Syriac, like the Hebrew, says *faces-of abyss,* not faces of-abyss, *i. e.* the first of two words in the genitive relation suffers change and not the second. §§ 76, 96 A.

18. The sign of the feminine gender is the letter ܬ. §§ 43. A, 76. 2.

19. The preformative ܢ marks the 3rd person of the Imperfect, § 45.

20. Roots have three letters (comp. ܐܡܪ), all other letters being preformatives or sufformatives, § 40.

21. When a Kushoy is over a letter preceded by a vowel, that letter is to be doubled, § 10. 2. (2).

22. The doubling of letters other than aspirates is not denoted by any sign, § 10. 2. (4).

23. The definite or emphatic state is denoted by the affix ܐ § 76, *Rem.* 1.

24. The plural is distinguished from the singular by Rebbuy, § 13.

25. Nouns have two numbers and two genders and three states § 76.

3. Grammar Lesson.

§§ 5, 16, 24, 33, 38, 39, 40. (25. 34 10, 6, 72. 93. 13. 17.
Review §§ 1—4. 34, 76.

4. Word Lesson.

ܡܳܐ *when.* ܐܬܪܚܦ *he brooded.*
ܢܦܰܩ *he went out.* ܚܙܳܐ *he saw.*

ܡܰܢ *who?* ܕ *that.*

ܝܺܕܰܥ *he knew.*

Exercises.

1. In beginning who made the earth? 2. Darkness brooded upon the earth, when God created the heavens. 3. Desolation was upon the faces of the heaven. 4. Darkness went out upon the waters and upon the faces of the earth. 5. The spirit of him who is God (was) brooding upon the waters. 6. He saw and knew that God created the heaven and the waters.

7. Write in Parallel columns and compare the Hebrew Perfect form כָּתַב and the Syriac form ܟܬܰܒ as to aspiration, vowels, the half vowel and the formatic elements.

8. Translate literally into English (or Hebrew) the Syriac of the lesson and retranslate.

LESSON THREE. Gen. I. 3. 4.

1. Notes.

22. ܘܰܚܙܳܐ—*wa-ḥ'zo, and (he) saw.*

(1) Compare 18 and 21.

(2) ܙ is Zain, a new letter.

23. ܠܢܽܘܗܪܳܐ—*l^{e}nuh-ro, the light.*

(1) ܠ is an inseparable preposition, § 34. It is often employed to denote the indirect object of the verb, § 123. It is also used in its true prepositional sense to denote the indirect object, § 124.

(2) ܢܽܘܗܪܳܐ. See 20.

24. ܕܫܰܦܺܝܪ—*dh^{e}shap-pir, that (it was) good.*

(1) ܕ is here a relative conjunction.

(2) ܕ=*dh* after a word ending in a vowel, § 10. 1. (3).

(3) ܫܰܦܺܝܪ is a masc. adjective in the absolute state.

(4) For the form of the noun, see § 72. 2. (4).

(5) The clause is an objective clause, § 125. 2, and a nominal sentence without a copula, § 117. 1.

25. ܘܰܦܪܰܫ—*wa-ph^{e}rash* (two syllables), *and (he) separated.*

(1) The first ܼ is a helping vowel, § 33. 2.

(2) The first syllable is open, § 17. 1.

(3) ܒ is pronounced with a half vowel (vocal shᵉwa), § 31. 1. 3.

26. ܒܶܝܬ݂ —*bhêth.—between.*

(1) ܒ݂ *bh* after a word ending in a vowel sound § 10. 1. (3).

(2) ܶܝ written fully, § 6. 5. (4).

(3) ܬ݂ *th* after a vowel, § 10. 1. (1).

27. ܠ—ܒܶܝܬ݂ *between—to.* Compare the Hebrew construction in vs. 6.

28. ܢܽܘܗܪܳܐ (see 12.).

29. ܘܩܪܳܐ—*wa-kᵉra* (two syllables) *and he called.*

(1) ܩܪܳܐ (compare 21) is in the simple verb stem 3rd person sing. masc., § 41. 1, § 43.

(2) The Imperfect would be ܢܶܩܪܶܐ *he shall call.* Compare ܢܶܗܘܶܐ *he shall be* 19.

30. ܠܢܽܘܗܪܳܐ—*lᵉnuh-ra* (two syllables), *to the light.*

(1) Notice that ܠ may denote the indirect object as well as the direct (comp. 23), § 124.

(2) ܠ is regularly prefixed without a vowel, § 34.

(3) ܗ ends the syllable and in Hebrew would have the silent Shᵉwa or syllable divider.—

31. ܐܺܝܡܳܡܳܐ—*'i-mama* (three syllables), *day.*

(1) Olaph is frequently placed before words for the sake of euphony. Before ܝ this Olaph takes ܺ in which the ܝ quiesces, § 20, *Rem.* 2.

(2) ܐܺܝܡܳܡܳܐ is day as opposed to night; ܝܰܘܡܳܐ at the end of the verse is the day of 24 hours.

32. ܘܠܚܶܫܽܘܟ݂ܳܐ—*wa-lᵉhesh-shu-kha* (four syllables, one for each vowel), *and to the darkness.*

Since three consonants can not come together at the beginning of a syllable, ܘ takes a helping Pethoḥo, § 33. 2.

33. ܩܪܳܐ (see 29) "*he called*".

34. ܠܺܠܝܳܐ—*l l-ya* (from *lai-lᵉyâ*), *night.*

(1) The emphatic ending has here lost its force, § 93. 1.

(2) ܺ is a contraction from *ai*, § 29. 3 (1).

35. ܪܰܡܫܳܐ—*ram-shâ, evening.*

(1) Emphatic state, the absolute and construct of which is ܪܡܫ.

(2) Though emphatic in form it is indefinite. (See 34).

36. ܨܰܦܪܳܐ *tsaph-râ, morning.*

37. ܝܰܘܡܳܐ ܚܰܕ *day one*, §§ 99. 1, *Rem.* 2, 93. I. 2, 93. II. 3.

(1) ܝܰܘܡܳܐ is indefinite in meaning being a translation of יוֹם.

(2) ܚܰܕ=אֶחָד the א being rejected, § 23. 1. (1).

(3) ܚܰܕ is in the absolute state agreeing with ܝܰܘܡܳܐ which though emphatic in form is absolute in meaning.

2. Observations.

26. Occurrence of aspirates in Gen. 1:1—5.

ܒܪܺܫܺܝܬ ܒ (*b*) follows nothing: ܬ (*th*) after ܝ.

ܒܪܳܐ ܒ (*b*) after ܬ of the preceding word.

ܝܳܬ ܬ (*th*) after the vowel ܳ.

ܗܘܳܬ ܬ (*th*) after the vowel ܳ.

ܬܽܘܗ ܬ (*t*) after ܬ of the preceding word.

ܘܒܽܘܗ ܒ (*bh*) after a half vowel sound.

ܘܚܶܫܽܘܟܳܐ ܟ (*kh*) after the vowel ܽܘ.

ܐܰܦܰܝ ܦ (*pp*) for *np*.

ܬܗܽܘܡܳܐ ܬ (*t*) after a diphthong.

ܕܰܐܠܳܗܳܐ ܕ (*d*) after a consonant, ܗ.

ܡܪܰܚܦܳܐ ܦ (*ph*) after a half vowel.

ܕܡܰܝܳܐ ܕ (*dh*) after ܐ of the preceding word.

ܦ (*pp*) the nominal form having the second radical doubled.

ܘܦܪܰܫ ܦ (*ph*) after the vowel ܰ.

ܒܶܝܬ ܒ (*bh*) after ܐ of the preceding word; ܬ (*th*) after ܶ.

ܨܰܦܪܳܐ ܦ (*ph*) after the vowel ܰ.

27. Six letters are called aspirates namely, ܒ, ܓ, ܕ, ܟ, ܦ and ܬ. With a point below these signs represent *bh* (=*v*), *gh*, *dh*, *kh*, *ph* and *th*; with a point above and when preceded by a consonant (*i. e.* neither vowel nor half vowel), they represent *b, g, d, k, p, t;* with a point above and when preceded by a vowel in the same word, they represent *bb, gg, dd, kk, pp, tt.*

28. There is no sign of doubling except for the aspirates.

29. There is no sign for Shᵉwa in Syriac.

30. The emphatic ending and state have often lost their force. See Notes 34 and 35 and § 93. 2.

31. This lesson has three new letters ܙ (*z*), ܩ (*ḳ*), ܨ (*ts*).

32. Notice the three states in:—

(1) Absolute: ܐܢܫ, ܬܘܗ, ܒܘܗ, ܫܡܝܢ.

(2) Emphatic: ܐܠܗܐ, ܫܡܝܐ, ܐܪܥܐ, ܫܡܫܐ, ܬܗܘܡܐ, ܢܘܗܪܐ, ܚܫܘܟܐ, ܟܠܟܠܐ, ܪܡܫܐ, ܨܦܪܐ, ܝܘܡܐ.

(3) Construct: ܐܦܝ.

33. Forms for special study: ܘܐܡܪ, ܘܗܘܐ, ܡܪܚܦܐ, ܦܪܫ.

3. Grammar Lesson.

(1) §§ 6, 23, 20, 41, 42, 43, 37.

(2) Learn the Imperfect of ܟܬܒ *to write* (§ 45).

(3) Review §§ 5, 9, 10, 11, 34, 40.

4. Word Lesson.

ܗܝܕܝܢ *then.*	ܣܡ *to put.*
ܙܒܢ *to buy.*	ܫܠܡ *to be finished.*
ܢܦܠ *to fall.*	ܕܠܚ *to be troubled.*
ܝܬܒ *to sit.*	ܩܪܒ *to be nigh.*
ܫܠܒ *to be at peace.*	ܣܓܕ *to worship.*

5. Exercises.

1. God said: let there be evening and let there be morning. 2. God shall call the morning light. 3. God saw the heavens and the earth and the waters and the good light (the light, the good). 4. In the beginning (was) darkness, then God made the light and he divided between the light and (to) the darkness.

5. Translate literally from Gen. I. 1—4 into Syriac and then retranslate.

6. Write the following forms in Syriac, giving Rukhokh and Kushoy. (1) Thou shalt write. (2) We shall write. (3) He shall write. (4) Ye

shall write. (5) I shall write. (6) We shall buy. (7) They shall buy. (8) Let him fall. (9) She sat. (10) He put. (11) It shall be finished. (12) Thou shalt be troubled. (13) Ye shall be nigh. (14) We shall worship. 15. Let it be at peace.

LESSON FOUR. Gen. I. 6—8.

1. Notes.

38. ܘܶܐܡܰܪ *and (he) said.*

(1) Syllables: (*a*) open, (*b*) closed, § 17. 1, 2.

(2) Vowels: (*a*) Rebhoṣo, (*b*) Pethoho, § 6.

(3) ܐ has here no consonantal force, but is quiescent, § 25.

(4) Pê Olaph verbs have a helping vowel with the Olaph, §§ 33. 1, 55. 1. This vowel when preceded by an inseparable particle is thrown back, the Olaph quiescing, §§ 25. 1. (2), 34. 2.

39. ܐܰܠܳܗܳܐ *God* (N. 3). (1) Three syllables, all open.

(2) Vowels: (1) Pethoḥo, (2, 3) Zeḳopho, § 6.

(3) The vowel ܰ has been retained by the guttural, as a helping vowel, § 28. 2. (2).

40. ܢܶܗܘܶܐ *neh-wê, let there be.*

(1) ܢ is the sign of the 3rd masc. of the Imperfect.

(2) Verbs whose last radical was originally ܘ or ܝ end in the Imperf. in ܐ ܶ § 60. 3.

41. ܪܩܺܝܥܳܐ (רָקִיעַ) *expanse.*

(1) Syllables: both open.

(2) Vowels: (1) Ḥebhoṣo (2) Zeḳopho.

42. ܒܡܨܰܥܬܳܐ *in the midst of.*

(1) ܒ has Rukhokh because the preceding word ends in a vowel; ܬ because preceded by a vowel, § 10. 1. (1). (3).

(2) ܒ is the inseparable preposition *in*, § 34. 1.

(3) ܡܨܰܥܬ is in the construct state singular number, fem. gender, § 76. 2.

43. ܡܰܝܳܐ *mă-yo, the waters.*

(1) The marks ̈ are Rebbuy, § 13, the sign of the plural.

(2) Some plurals end in ܐ ܶ, § 86. 16.

44. ܦܳܪܶܫ ܢܶܗܘܶܐ *let it be dividing.*

(1) ܢܶܗܘܶܐ is the same as in N. 38. Used with a participle, it denotes continuous future action, § 127. 5.

(2) ܦܳܪܶܫ—*po-resh* is a participle of the simple species. Compare פֹּרֵשׁ. See § 50.

(3) The original form of the participle was *pârĭsh*; the ô from â is naturally long and hence unchangeable, the ĭ is naturally short and changeable, § 7. 3.

45. ܒܶܝܬ—*bêth, between.*

(1) ê is naturally long by contraction from *ăy*, §§ 7. 3, 29. 3.

(2) The root is ܒܰܝܢ; *n* has been dropped (as in בַּת in Hebrew). The form was ܒܰܝܢܬ, § 23. 2. (3).

46. ܘܰܥܒܰܕ—*wă·ʿebhădh, and he made.*

(1) ܥ is to be distinguished form ܠ, § 4. 3. (6).

(2) ܘ takes a helping vowel § 34. 1, and forms with it a half open syllable, § 17. 4.

(3) ܥ has a half-vowel as is shown by the aspirated ܒ, §§ 9. 3, 10. 1. (2).

(4) ܥܒܰܕ is the 3rd pers. sing. masc. of the Peʿal or simple form of the verb. This is the *first form* of the verb and the simplest, § 43. 1.

47. ܐܰܪܩܺܝܥܳܐ—*'arkîʿo, the firmament.*

By comparison with 39, it will be seen that an Olaph has been placed before ܪܩܺܝܥܳܐ. This Olaph is called Olaph prosthetic, § 20, 1.

48. ܘܦܰܪܶܫ and he separated.

This is in the simple or Peʿal stem, in the first form, see 46.

49. ܕܰܠܬܰܚܬ—*dalethah'*, *which (were) beneath* (lit. *to under*).

(1) ܬܰܚܬ *under*, is a preposition.

(2) ܠ *to* is an inseparable preposition, § 34.

(3) ܕ is the relative pronoun. It here introduces the relative or adjective clause defining ܡܰܝܳܐ, §§ 38, 136.

50. ܡܶܢ *from*, see § 6. 3. (2).

51. ܕܰܠܥܶܠ—*dalᵉ·el, which were above.*

(1) Three elements: the relative ܕ cf. 49. 3, the preposition ܠ, see 49. 2, and ܥܰܠ = ܥܶܠ *upon*, see 13.

(2) For the change from ܥܶܠ to ܥܰܠ see § 29. 2.

52. ܘܰܗܘܳܐ *and it was.*

(1) ܗܘܳܐ is the first form of the verb.

(2) The ܳ comes from original *awa*, § 60. 1, but see also § 29. 5. (2), (3).

53. ܗܳܟܰܢܳܐ—*hokhanno, so.*

54. ܕܰܬܪܶܝܢ—*dathᵉrên, which is two* i. e. *second.*

(1) ܕ, pointed according to § 34 with a helping vowel, is the relative pronoun introducing an adjective clause, § 38.

(2) ܬܪܶܝܢ is one of the few remnants of a dual which remain in Syriac, § 76. 5.

(3) Notice that Hebrew שׁ often becomes ܬ in Syriac and ז is sometimes changed to ܕ.

2. Observations.

34. A helping vowel is given to every second consonant preceding one which has a vowel, e. g. ܘܰܟܬܰܒ, ܘܰܗܘܳܐ, ܕܰܬܪܶܝܢ, ܕܰܠܥܶܠ.

35. The vowel of prosthetic Olaph or of an Olaph beginning a word is thrown back upon a previous ܘ, ܕ or inseparable preposition, e. g. ܘܶܐܡܰܪ, ܠܰܐܪܩܺܝܥܳܐ.

36. A word has as many syllables as it has full vowels, e. g. ܘܶܐܡܰܪ *we-mar*, ܐܰܠܳܗܳܐ *'a-lo-ho*, ܪܩܺܝܥܳܐ *rᵉḳi-ʿo.*

37. Pretonic *ā* is not found in Syriac. The original *ă* of the verbal or nominal form, which in Hebrew becomes pretonic Kamets, is volatilized, e. g. ܟܬܰܒ = כָּתַב.

38. The feminine in Syriac is everywhere denoted by ܬ, except in the absolute state of nouns and participles, e. g. ܪܚܺܝܡܬ, ܗܘܳܬ, ܡܰܠܟܬܳܐ but ܡܰܠܟܳܐ. Such nouns as ܐܰܪܥܳܐ and ܢܰܦܫ are, in the singular, masculine in form but feminine in gender. Compare the agreement of the nouns with feminine verbs and see § 86. 10.

39. The name of the simple species or stem is Pᵉʿal. Examples of it are: ܦܪܰܫ; ܟܬܰܒ and ܗܘܳܐ.

40. The 3rd pers. sing. masc. of the Pᵉʿal is the first as well as simplest form of the verb, from which all other forms of different gender number person tense or stem are made by vowel changes and by significant preformatives and sufformatives; and, in the case of the intensive stems, by the doubling, also, of the second radical.

41. Naturally long vowels are unchangeable e. g. ô from original â ܦܪܫ, î in ܢܦܝܚܐ, ô in ܗܘܬ, û in ܪܘܚܗ, ê in ܫܡܐ.

42. Naturally short vowels are changeable, e. g. ܡܪܚܦ, ܡܪܚܦܐ, ܦܪܫ, ܦܪܫܐ.

3. Grammar Lesson.

1. §§ 7. 14—17. 25. 30. 31. 35. 45.

2. Learn the following table which gives the regular inflection of substantives, adjectives and participles.

	Sing. masc.	Sing. fem.	Plur. masc.	Plur. fem.
Abs.	ܦܪܫ	ܦܪܫܐ	ܦܪܫܝܢ	ܦܪܫܢ
Cons.	ܦܪܫ	ܦܪܫܬ	ܦܪܫܝ	ܦܪܫܬ
Emph.	ܦܪܫܐ	ܦܪܫܬܐ	ܦܪܫܐ	ܦܪܫܬܐ

3. Show what the endings for gender and number are.

4. What vowel is unchangeable? What vowel is everywhere dropped before forms with affixes?

5. Pronounce *por-sho* not *porᵉsho*; *por-shîn* not *porᵉshîn*, §§ 30. 1, 31. 3. Rem. 2.

6. Inflect, in like manner, ܡܪܚܦ *brooding*, § 81.

7. Review §§ 5. 6. 16. 24. 33.

4. Word Lesson.

ܒܝܫ *evil.* ܕ (inseparable prefix) *which, that.* ܝܡܐ *sea.* ܫܠܚ *he sent.* ܬܪܝܢ *two.* ܡܢ *from.* ܗܘ *it, he.* ܬܠܡܝܕ *disciple.*

5. Exercises.

1. Let there be the firmament between the waters and the waters. 2. Between the heavens and the earth which God made. 3. The

waters which are under from the heavens and the earth which is above from the seas. 4. In the morning which is two God sent the light and he called the light day. 5. It (is) good that it was so. 6. The disciples the good [are] separating from the evil. 7. God [is] good and the earth (fem.) which (ܕ) he made [is] good. 8. In the beginning the light (was) separating between the day and the night.

9. Translate literally verses 6 and 7 and, without further aid than your own translation, translate back into Syriac.

10. Do the same for verses 1—5.

LESSON FIVE. Gen. I. 9—13.

1. Notes.

55. ܢܶܬ݂ܟܰܢܫܽܘܢ—*nethkanneshun, let them be collected.*

(1) The preformative ܢ is the sign of the 3rd person of the Impf.

(2) The ܽܘܢ is the sign of the masculin plural.

(3) The ܬ prefixed after a preformative is the sign of the reflexive or passive stem.

(4) The root is ܟܢܰܫ *to collect.*

(5) This form would be written the same in the Ethpe'el and in the Ethpa'al. It is better here to put it in the latter, or the intensive passive; and to read *neth-kan-neshun,* not *neth-ka-neshun.*

56. ܠܰܐܬ݂ܪܳܐ—*lath-ro, to a place.*

(1) The vowel of the Olaph is thrown back to the inseparable preposition, § 34. 2.

(2) Though emphatic in state, the noun is indefinite in meaning, § 93. 2.

57. ܬܶܬ݂ܚܙܶܐ—*teth-ḥezê, let appear.*

(1) The preformative ܬ shows that the form is in either the 3rd fem. sg. or in the 2nd. masc. The context decides for the former.

(2) The first form of all the Imperfects of verbs Lomadh Hê ends in ܶܐ. The 3rd fem. sg. differs from the masc. merely in changing ܢ to ܬ, § 45. 5.

(3) The first form of the verb is ܚܙܳܐ, see 22, and compare ܗܘܳܐ, ܗܘܳܐ and ܢܶܗܘܶܐ.

(4) The second ܬ shows that this is a passive stem, § 41. 4. Since the Ethpa'al and Ettaphal have everywhere three syllables, this must be the passive of the simple stem, i. e. the Ethpe'el.

58. ܝܰܒܺܝܫܬܳܐ—*yab-bîsh-to, the dry land.*

(1) The ܒ being preceded by a vowel the dot above it, called Kushoy, shows that it is to be doubled, § 10. 2. (2).

(2) ܬ not having a vowel before it, the dot shows merely that there is no half-vowel after the ܫ and that ܬ is unasperated.

(3) ܬ is the sign of the fem.; ܐ of the emphatic state, § 76.

59. ܘܰܠܟܶܢܫܳܐ—*wal^ekhensho, and to the gathering.*

(1) ܟܶܢܫܳܐ is a segholate noun masc. sing. emphat. § 67.

(2) ܠ is the inseparable preposition regularly prefixed, § 34.

(3) ܘ is the conjunction. It has a helping vowel because occurring before an unvowelled consonant. It forms with this vowel a half-open syllable and the ܠ takes a half-vowel., §§ 17. 4, 31. 3, 33. 2.

60. ܕܡܰܝܳܐ—*d^emayo, of the waters,* lit. *that of the waters.*

(1) ܕ is really a demonstrative pronoun corresponding to Hebrew זֶה. It has come to denote the genitive relation, § 98. A.

(2) For ܡܰܝܳܐ, see 17.

61. ܝܰܡܡ̈ܶܐ—*ya-mê, seas.*

(1) The two dots are the sign of the plural *Rebbuy*, § 13.

(2) The line under the first ܡ is linea occultans and shows that, although written the ܡ is not to be pronounced, § 19. 3.

(3) ܐ is the sign of the emphat. masc. plural., § 76. 3.

(4) The singular is ܝܰܡܳܐ, § 67. 2. (7).

62. ܬܰܦܶܩ—*thap-pek, let cause to go out.*

(1) ܬ with Rukhokh because the word preceding ends in a vowel, § 10. 1. (3); ܦ with Kushoy, because doubled to compensate for an assimilated Nun, § 10. 2. (2).

(2) The ܬ shows the 3rd fem. Impf.; the ܰ above it is the sign of the causative or Aph'el stem, §§ 45. 2, 47. *Rem.* 4.

(3) The Yudh at the end is sometimes, though less seldom than not, found with the 3rd fem. Impf. § 47. *Rem.* 5, § 45. 5.

(4) The root is ܢܦܩ, the Nun being assimilated always at the end of a syllable when not accompanied by a vowel, §§ 18, 53.

63. ܬܕܐܐ—*tha-dho, grass.*

(1) The last Olaph is otiant, § 24. 1; the other is quiescent, its vowel having been thrown back on the ܕ, so that we have *tha-dho* instead of *thadh-'o*, § 25. 1. (2).

(2) ܬ has Rukhokh because the preceding word ends in a vowel, § 10. 1. (3).

(3) ܬ stands for Hebrew שׁ. It has been transposed with ܕ, the vowel being equivalent to דֶּשֶׁא

64. ܥܣܒܐ—*'es-bo, herb.*

(1) This is a segholate of the ĭ class, § 67. 1.

(2) It is in the emphatic state of the masc. although indefinite § 93. 2.

65. ܕܡܙܕܪܥ—*dh^emez-d^era', which was seeding for itself.*

(1) ܕ is the relative pronoun and introduces the adjective clause, §§ 38. 136. It has Rukhokh after a preceding vowel, § 10. 1. (3).

(2) ܡܙܕܪܥ is the Ethp^e'el or reflexive of the simple stem. ܡ is the sign of the participle, § 50. 2. ܕ and ܙ have been transposed, according to § 21. 1. ܕ is metathesis from ܬ the sign of the reflexive, § 22. 4. In the last syllable the vowel is ̱ instead of ̄ because of the guttural, §§ 26. 1. (1), 52. 3.

66. ܠܓܢܣܗ—*l^egen-seh, according to its kind.*

(1) ܠ is the inseparable preposition regularly prefixed with a half-vowel, § 34. 1.

(2) ܗ̄ is the pronominal suffix 3rd masc. sing. (=הוּ), § 36.

(3) ܓܢܣ is treated as an ă class segholate. It comes from the Greek γένος.

67. ܘܐܝܠܢܐ—*wî-lo-no, and the tree.*

(1) Waw draws back the vowel of the ܐ and the Olaph quiesces, § 25. *Rem.* 1.

(2) ܐ is the sign of the emphatic state, § 76. *Rem.* 1.

68. ܕܦܐܖ̈ܐ—*dh^ephî-rê, of the fruit,* lit. *that of the fruits.*

(1) ܕ is aspirated after the preceding vowel. It introduces an

appositional relative clause which has become equivalent to our genitive, § 97. A. 2.

(2) ܦܐܖ̈ܐ. The two dots are Rebbuy, one of them standing also for the diacritical point of the Rish, § 13. 2. ܐ̈ is the sign of the masc. plur. emph., § 76. 3.

69. ܕܥܳܒܶܕ—*dhe'o-bhedh, which was yielding*, lit. *was making.*

(1) For ܕ see 65. 1.

(2) ܥܳܒܶܕ is the active part. of Pe'al. Masc. sing., see 44.

70. ܕܢܶܨܒ݂ܬܶܗ—*denes-betheh, whose stock.*

(1) ܕ introduces the relative or adjective clause. It has Kushoy because preceded by a consonant. Along with ܗ his it forms the genitive *of which* or *whose*, §§ 36, 38, 104. 2.

(2) ܢܨܒܬ is in the fem. as shown by ܬ § 76. 2. The emphatic is ܢܨܒܬܐ (cf. Heb. מַצֶּבֶת Is. 6:13).

71. ܒܗ *in it.*

This is the inseparable preposition ܒ and the masc. sing. 3rd pers. pron. suffix. §§ 34, 36. 3.

72. ܘܐܰܦܶܩܰܬ—*wap-pekath, and (she) caused to go forth.*

(1) Wau, as usual, draws back the vowel of the Olaph the latter quiescing, §§ 25. *Rem.* 1, 34. 2.

(2) In ܐܰܦܶܩܰܬ, ܰܬ is the sign of the 3rd fem. sing. of the perfect; ܐܰ is the sign of the Aph'el or causative stem; the Kushoy over the ܦ, since it is preceded by a vowel, shows that the ܦ is doubled; the doubling is occasioned by a preceding Nun, which has been assimilated regularly at the end of a syllable when preceded by a vowel and followed by none, §§ 43. *Rem.* 1, 41. 3, 53. 2.

73. ܕܬܠܳܬܳܐ—*dhathelo-tho, which is three* i. e. *the third.*

(1) This is an adjective clause limiting ܝܰܘܡܳܐ *day.*

(2) In *dha, dh* is aspirated after the preceding vowel; *a* is a helping vowel; the syllable is half-open, §§ 33. 2, 17, 4.

(3) ܬܠܳܬ is equivalent to the Hebrew שָׁלֹשׁ, שׁ being regularly equivalent to ܬ where they both correspond to ث in Arabic. The pretonic Kamets of the Hebrew is always volatilized in Syriac.

2. Observations.

43. There are in Syriac:

(1) A simple verb stem, *e. g.* ܚܙܳܐ, ܦܪܰܩ.

(2) An intensive verb stem, *e. g.* ܩܰܪܶܒ.

(3) A causative verb stem, *e. g.* ܐܰܩܛܶܠ.

(4) A simple passive stem, *e. g.* ܐܶܬܚܙܺܝ, ܐܶܬܦܪܶܩ.

(5) An intensive passive stem, *e. g.* ܢܶܬܩܰܛܠܽܘܢ.

(6) And a causative passive stem, (not yet occurring).

44. The characteristic of the intensive stems is the doubling of the second radical.

45. The causative stem is characterized by ܰ before its first radical.

46. All passives have as their sign a ܬ occurring before the first radical. In all Perfects and Imperatives this ܬ is preceded by Olaph; in all Participles and Infinitives by Mim; in the Imperfect by the appropriate personal preformative.

47. The names of the stems are P^e'al, Pa'el, Aph'el, Ethp^e'el, Ethpa'al, Ettaph'al.

48. ܳ when naturally long corresponds to the Hebrew naturally long *ô*.

49. The name of ܳ is Z^{e}ḳopho; of ܰ, P^{e}thoḥo; of ܶ R^{e}bhoṣo; of ܺ Ḥ^{e}bhoṣo; of ܽ '^{e}ṣoṣo.

50. The preformatives of the Imperfect are the same as in Hebrew, except that in the 3rd person masculine there is Nun instead of Yudh and that in the 3rd fem. plur. there is Nun instead of Tau.

3. Grammar Lesson.

(1) Review the sufformatives of the P^e'al Perfect, § 43.

(2) Form with the aid of these the Perfects of all the other stems, § 44.

(3) §§ 8, 13, 18, 19, 21, 22, 36, 44.

(4) Review §§ 9—11, 20, 23, 34, 35, 37—43.

4. Word Lesson.

ܢܣܰܒ *to take.*	ܫܡܰܥ *to hear.*
ܐܰܣܺܝ *to heal.*	ܙܩܰܦ *to crucify.*

ܠܒܫ *to clothe.*	ܐܝܟ ܕ *according as.*
ܐܬܢܣܝ *to tempt.*	ܕ *what, that.*
ܦܩܕ *to command.*	ܠܡܐܡܪ *to say.*
ܩܪܒ *to be near.*	ܡܛܠ ܕ *because.*
ܪܚܩ *to be far.*	ܫܪܪܐ *truth.*
ܦܬܚ *to open.*	ܐܢܘܢ *them.*

5. Exercises.

1. God said: Let the heavens be opened and let the herb appear on the earth. 2. The earth brought forth grass according as God commanded. 3. Let the earth bring forth the tree of fruits which is making fruits whose sprout is in itself. 4. He clothed the earth (with) herbs. 5. Hear ye what I have been commanded to say. 6. Ye have been healed because ye have heard what I said. 7. He was taken and tempted and crucified. 8. Ye have been commanded to say the truth. 9. God clothed them and commanded them to hear the truth. 10. Draw nigh to God and He will draw near to you, remain far from Him and He will be far from you.

LESSON SIX. Gen. I. 14—16.

1. Notes.

74. ܢܶܗܘܽܘܢ—*neh-wun, let them be.*

(1) The first Nun indicates the Imperfect 3rd person, § 45.

(2) The ending ܘܢ denotes the masculine plural, § 45. 6.

(3) The root is ܗܘܐ *he was,* cf. ܗܘܬ *she was.* See 9.

75. ܢܰܗܺܝܪ̈ܐ—*nah-hîrê, lights.*

(1) The two dots over the Rish are Rebbuy, the sign of the plural; one dot coincides with the diacritical point of the Rish, § 13. 2.

(2) ܐ is the sign of the masc. plur. emphatic, § 76. 3.

(3) The root is the same as that of ܢܘܗܪܐ *light,* see 20.

76. ܕܰܫܡܰܝܳܐ—*dha-sh^ema-yo, of the heavens,* lit. *that of the heavens.*

(1) ܕ is in apposition with ܐܪܩܝܥܐ; ܫܡܝܐ is in the genitive relation to the pronoun. See § 97. A.

(2) The first syllable is half open and its vowel a helping vowel, §§ 17. 4, 33. 2.

77. ܠܡܶܦܪܰܫ—*l^emeph-rash, to separate.*

(1) ܠ is the inseparable preposition regularly prefixed with a half-vowel, § 34. 1. It is always used before the Infinitive construct § 120. 1. (3).

(2) ܡܶܦܪܰܫ is the Infinitive of the Pe'al, § 49. 1. The root is ܦܪܰܫ.

78. ܡܶܫܬܰܥܢܐ, cf. ܐܶܫܬܰܥܢܐ see 31. In one case there is Olaph prosthetic; in the other not. Cf. ܪܩܝܥܐ 41 and ܐܪܩܝܥܐ 47.

79. ܠܺܠܝܐ—*li-l^eyo, night.*

(1) *i* comes by contraction from *ay*, § 29. 4. (4).

(2) This noun is generally written ܠܠܝܐ. See 41.

80. ܠܳܬ̈ܘܳܬܐ—*loth^ewotho, for signs.*

(1) ܠ is the preposition, which draws back the vowel of the Olaph, the latter quiescing, § 25. 1. (2) and *Rem.* 1.

(2) ܬܐ̈ is the sign of the fem. plur. emphatic. The Wau is sometimes inserted in nouns between the root and the ending, § 86. 3.

(3) The singular is ܐܳܬܐ.

81. ܘܰܠܙܰܒܢ̈ܐ—*wa-l^ezabh-nê, and for times.*

(1) When more than one of the inseparable particles occur together every second one takes a helping vowel, § 34. 4.

(2) The noun is masc. plur. emph.; see 75. 2.

82. ܘܰܠܝܰܘܡ̈ܳܬܐ—*wa-l^eyau-mo-tho, and for days.*

(1) ܘܰܠ as in 81. 1.

(2) ܬܐ̈ fem. plur. emphatic see 80. 2. The two dots are Rebbuy. The singular is ܝܰܘܡܐ.

83. ܘܠܰܫ̈ܢܰܝܐ—*w^ela-sh^ena-yo, and for years.*

(1) Since Shin is without a vowel Lomadh takes a helping vowel and Wau does not. Cf. 81 and 82.

(2) ܫ̈ܢܰܝܐ is a plur. emphatic. See § 86. 16. The singular is ܫܰܢ̱ܬܐ, § 87. 30.

84. ܡܰܢܗܪܺܝܢ—*man-h^erîn, giving light.*

(1) ܡ prefixed is a sign of participle except in Pe'al. Cf. ܡܦܰܪܣܐ (16) and ܡܬܪܰܚܦܐ (65), but ܦܳܪܶܫ (44. 2.) and ܚܳܫܶܟ (69).

(2) P^e^thoḥo with the preformative of the Participle denotes the causative or Aph'el stem, § 41. 3.

(3) ܝܢ is the sign of the masc. plur. absolute, § 76. 3.

85. ܠܡܢܗܪܘ *l^e^man-ho-ru, to give light.*

(1) ܡ is prefixed to all Infinitives. The Infinitive construct is always preceded by ܠ, §§ 49, 120.

(2) All Infinitives, except the P^e^'al end in ܘ with before the last radical, § 49. 2.

(3) P^e^thoḥo with the preformative denotes the causative stem. Cf. 84. 2 and see § 41. 3.

86. ܬܪܝܢ *th^e^rên, two.*

(1) For the etymology, see § 76. 5.

(2) For the syntax, see § 110. A.

87. ܪܘܪܒܐ—*rau-r^e^bhê, great.*

(1) This is an irregular plural from ܪܒ, in the emph. state, § 87. 27.

(2) Note the position of the adjective after its noun and its agreement in gender, number and state. The same is true of ܪܒܐ *great* and ܙܥܘܪܐ *small*, §§ 93. 3. (1), 99. 1.

88. ܕܐܝܡܡܐ—*dhîmomo, of the day.*

(1) This is the second kind of the genitive constructions, § 97. A.

(2) ܕ is aspirated according to, § 10. 1. (3).

(3) Olaph prosthetic quiesces in the which has been thrown back on the preceding ܕ, § 34. 2, § 20, *Rem.* 2.

89. ܘܟܘܟܒܐ—*w^e^khau-k^e^bhê, and the stars.*

(1) The first Kaph has Rukhokh after a half-vowel; the second has Kushoy after a diphthong, § 10.

(2) ܐ is the sign of the masc. plur. emph.; the two dots are Rebbuy, §§ 13, 76. 3.

2. Observations.

51. All Infinitives have the preformative ܡ.

52. We have had, so far, three ways of denoting the genitive relationship.

(1) ܡܬܝܗܒܐ ܡܢܬܐ vs. 6.
ܐܦܩܬ ܠܬܘܗܡܐ vs. 2.

(2) ܐܝܠܢܐ ܕܦܐܪܐ vs. 11.
ܐܘܪܚܐ ܕܫܡܝܐ vs. 14.

(3) ܢܘܗܪܗ ܕܚܒܪܗ vs. 2.

53. The Preformative of all Participles except the P'al, is ܡ. The only mark to distinguish the Infinitive from the first form of the Participle is the ending ܘ and the vowel ܳ before the last radical.

54. Participles have the inflection of nouns.

55. The preformatives of Infinitives and Participles have the same vowels as the Imperfect of their respective stems.

56. The preformative of the P'al stem is ܶ, *e. g.* ܢܦܪܘܩ 77, ܢܗܘܘܢ 74; the Pa'el stem has a half-vowel with the preformative, *e. g.* ܡܦܪܫܐ 16; the Aph'el has ܰ *e. g.* ܐܦܩܬ 72, ܢܥܒܪܘܢ 84. The preformatives of all the reflexive or passive stems are all followed by ܬ *e. g.* ܬܬܚܙܐ 57, ܬܬܟܢܫܘܢ 55, ܢܬܪܘܡܡ 65.

57. Attributive adjectives follow their nouns and agree with them in gender, number and definiteness.

3. Grammar.

1. Review the sufformatives and preformatives of the P'al Imperfect; and form with the aid of these the Imperfects of all the derived stems, §§ 45, 47.

2. §§ 12, 77.

3. Review §§ 76. 1—8. 13—18. 45.

4. Word List.

ܙܩܦ *to crucify.*
ܗܠܟ *to go.*
ܟܢܫ *to assemble.*
ܐܬܩܪܒ *to draw near.*
ܐܪܚܫ *to swarm.*
ܐܬܬܘܣܦ *to add.*
ܐܝܠܝܢ ܕ *those which.*
ܟܬܒ *to write.*
ܟܬܒܐ *a book.*
ܫܡܫܐ *sun.*
ܡܛܠ ܕ *because.*
ܡܢ *from.*
ܦܩܕ *to command.*
ܟܕ *when.*

5. Exercises.

1. The sun will be seen for the rule (that) of the day. 2. God made great stars for signs and for times. 3. The sun and the stars shall be shining in the expanse of heaven to give light upon the earth and they shall be the signs of the seasons of the days and of the years. 4. To divide; he shall divide; they shall divide; dividing; they divided; divided. 5. Crucify them; let them be crucified; thou shalt be crucified. 6. Thou shalt go in darkness because thou hast drawn near and hast taken from the fruits of (ܕ) the tree. 7. The waters (*pl.*) were assembled into one place and swarmed because God had so commanded. 8. Two books (two the books) shall be added to those which have been written and thou shalt write them when they shall be written.

LESSON SEVEN. Gen. 1. 17—23.

1. Notes.

90. ܘܝܗܒ—*w*^{e}*yabh, and (he) gave.*

(1) The line ander ܗ is linea occultans, § 11.

(2) ܝܗܒ is the P^{e}'al Perfect first form, see § 64. 7.

The usual first form for a Pê Yudh verb would be ܝܗܒ § 58.

91. ܐܢܘܢ—*'ennun, them.* There is no pronominal suffix for the 3rd plural with verbs. In its stead, the personal independent pronoun is used, § 36. 2.

92. ܠܡܫܠܛ—*l*^{e}*mesh-lat, to rule over.* This is the P^{e}'al Infin. from ܫܠܛ Cf. 77.

93. ܕܐܪܒܥܐ—*dharb*^{e}*'o, which is four.*

(1) This is a relative clause § 136.

(2) The cardinal after the relative may take the place of the ordinal § 110 B.

94. ܢܪܚܫܘܢ—*narḥ*^{e}*shun, let (them) swarm.*

(1) Nun is the preformative of the 3rd person Imperf. everywhere except in the 3rd fem. sing.

(2) ܘܢ is the ending of the masc. plural Imperf.

(3) P^ethoḥo with the preformative is the sign of the Aph'el stem, § 41. 3.

95. ܪܰܚܫܳܐ—*raḥ-sho, a swarm.*

(1) This is an *a* class segholate in the emphatic singular, § 67.

(2) Rebbuy shows that the noun is a collective, see § 90. 1.

96. ܚܰܝ̈ܬ݂ܳܐ—*ḥay-y^etho, living.*

(1) The Rukhokh under the Tau shows that the preceding Yudh is doubled; since if *ay* were a simple diphthong Tau would have Kushoy, § 10. 2. (3) *Rem.*

(2) ܬ݂ܐ is the sign of the fem. sing. emphatic, § 76. 2.

97. ܦܳܪܰܚܬ݂ܳܐ—*po-ra-ḥ^etho, bird.*

(1) The fem. abs. is ܦܳܪܚܳܐ—*po-r^eḥo*; in the emphatic the short vowel is retained in order to avoid the coming together of three consonants at the beginning of a syllable, §§ 16. 2, 33. 2.

(2) The participle is here used as a noun, § 118.

98. ܬܶܦܪܰܚܝ—*theph-raḥ, let (her) fly.*

(1) The preformative ܬ is the sign of the 3rd fem. sing. Imperf.

(2) The sufformative ܝ is sometimes found with the 3rd sing. Imperf. Cf. ܬܦܘܩܝ 62.

(3) Yudh is otiose, § 24. 3.

(4) The ܶ shows it is in the simple or P^e'al stem. Obs. 56.

(5) P^ethoḥo is the usual vowel over the 3rd radical in Lomadh Guttural verbs, § 52.

99. ܕܰܐܪܚܶܫܘ—*dhar-ḥesh, which (they) caused to swarm.*

(1) ܕ has Rukhokh according to § 10. 1. (3). It draws back the vowel of Olaph, § 34. 2. The Olaph quiesces in the preceding vowel, § 25. 1. (2).

(2) The Olaph designates the Aph'el stem; the Wau shows the 3rd plural, §§ 41. 3, 43. 6.

(3) The Wau is otiose, § 24. 2.

100. ܓܶܢܣܗܽܘܢ—*gen-s^ehun, their kind.*

(1) For ܓܶܢܣ, see 66. 3.

(2) ܗܽܘܢ is the pronominal suffix of the 3rd plur. masc. *with nouns.* It is never used with verbs, see 91 and §§ 36, 77.

101. ܕܓܶܦܳܐ—*dh'ghep-po, of wing.*

(1) The Kushoy in the Pê is by way of compensation for an assimilated Nun, §§ 10. 2. (2), 18. 1, 67. 2. (6).

(2) The root ܓܢܦ is cognate to כנף.

(3) ܓܶܦܳܐ (ܓܶܢܦܳܐ) is in the emph. sing. masc. Cf. ܢܰܦܫܳܐ, ܓܶܢܣܳܐ.

102. ܓܶܢܣܳܗ̇—*ghen-soh, her kind.*

(1) ܳܗ̇ is equivalent to the Hebrew הָ, § 36.

(2) The diacritical point denotes the fem. ̇ as distinguished from the masc. ̈, § 6. 6. (2).

(3) The fem. suffix refers back to ܦܳܪܰܚܬܳܐ. Cf. ܓܶܢܣܶܗ 66, ܓܶܢܣܗܽܘܢ 100.

103. ܒܰܪܶܟ—*bar-rekh, (he) blessed.*

(1) The ܰ over the first radical designates the intensive or Pa'el stem. Cf. ܡܪܰܚܦܳܐ, § 41. 2.

(2) This is the Perfect, since it has no preformative and cannot be in the Imv. It is in the first form *i. e.* 3rd masc. sing., since it has no sufformative for gender, number or person, § 43. 4.

104. ܠܗܽܘܢ—*l'hun, to them.*

(1) ܠ is a preposition. Cf. vs. 12.

(2) ܗܽܘܢ is pronominal suffix 3rd plural masc. See 100.

105. ܦܪܰܘ—*p'rau, be fruitful.* This the Imv. 2nd. masc. plur. of the simple stem from a Lomadh Olaph root, § 60. 4.

106. ܘܰܣܓܰܘ ܘܰܡܠܰܘ—*wa-s'ghau wa-m'lau, and multiply and fill.*

(1) The vowel with the Wau is a helping vowel, § 33. 2.

(2) The syllable after Wau is half-open, § 17. 1.

(3) These are both in the 2nd masc. plur. Imv. P'al.

107. ܕܰܒܝܰܡܡܶܐ—*dha-bh'ya-mê, which are in the seas.*

(1) The line under the first ܡ is linea occultans. See 61.

(2) The singular is ܝܰܡܳܐ.

(3) The clause is relative. § 136.

108. ܬܶܣܓܶܐ—*thes-gê, let (it) multiply.*

(1) The preformative ܬ denotes the 3rd sing. fem. Imperfect; the vowel *e* used with the preformative designates the simple stem.

(2) ܐ shows the root to be Lomadh-Olaph. Cf. ܬܶܗܘܶܐ 40. ܬܶܬܚܙܶܐ 57.

109. ܕܚܡܝܫܝܐ *the fifth*, see 93; ܬܦܩ *let bring forth*, see 98; ܪܚܫܐ *creeping thing*, see 96.

110. ܚܝܘܬܐ—*hay-wetho'*, *beast.*

(1) The line above the Wau is called Marhetono, and shows that Wau is to be pronounced without a vowel, § 12. 2.

(2) The noun is a fem. segholate of the *a* class, § 67. 3.

2. Observations.

58. The Infinitive construct is always preceded by ܠ.

59. The cardinals preceded by the relative are often used for the ordinals.

60. The inseparable prepositions take pronominal suffixes *e. g.* ܒܗ ܠܗܘܢ 104.

61. Nouns take pronominal suffixes *e. g.* ܠܓܢܣܗ 66, ܠܓܢܣܗ 102, ܠܓܢܣܗܘܢ 100.

62. The 3rd pers. masc. of the personal pronoun is not suffixed to the verb, see vs. 17, 22.

63. Final Wau and Yudh do not take the linea occultans, when unpronounced *e. g.* ܬܦܪܚ 98, ܕܐܪܚܫܘ 99.

64. Nun is the preformative of the 3rd pers. Imperf. everywhere except in the 3rd fem. sing. where we have Tau.

65. ܘܢ is the ending of the masc. plur. Imperfect; ܘ of the masc. plur. Perf. and Imv.

66. Collectives sometimes take Rebbuy, see 96 and § 90.

66*a*. At the end of a syllable, Nun is assimilated to the succeeding consonant, which is then doubled. See 101, 109.

3. Grammar Lesson.

(1) Personal pronouns and pronominal suffixes §§ 35, 36. 2.

(2) §§ 32, 48, 49, 50.

(3) ܝܗܒ *to give*, § 64. 7.

(4) Review, §§ 19—25.

4. Word Lesson.

ܫܰܠܶܛ *to rule.*
ܫܒܰܩ *to forsake.*
ܥܒܰܕ *to make.*
ܐܰܠܶܦ *to teach.*
ܥܡܰܕ *to baptize.*
ܐܰܢܗܰܪ *to shine.*
ܐܶܫܬܡܰܥ *to obey.*
ܟܰܕ *when.*
ܛܳܒ ܡܶܢ *better then.*

5. Exercises.

1. God gave the smaller light for the ruling of the night and the stars to shine upon the earth and to separate between the light and (to) the darkness. 2. He made the great lights to rule over the day and over the night and when he saw them he said that (it was) good. 3. God taught the birds (sing.) to fly upon the face of the firmament of heaven. 4. Every living soul was taught to obey God who created all and blessed them and said to them: Obey God and ye shall be blessed; forsake God and He will destroy the great seamonsters and every living soul which creepeth (Participle absolute fem. sing.) and every bird of wing which shall multiply in the earth. 5. Ye shall be baptized with water and with the Spirit, which brooded over the face of the waters when God created the heavens and the earth and all which was in them. 6. To obey is better than to rule.

LESSON EIGHT. Gen. I. 24—31.

1. Notes.

111. ܘܟܽܠܶܗ ܪܰܚܫܳܐ *and all of it* (to wit) *the creeping things.*

(1) ܗ̣ is a pronominal suffix agreeing in gender and number with the collective ܪܰܚܫܳܐ.

(2) ܪܰܚܫܳܐ is in apposition with ܗ̣.

(3) For the construction, see § 108. 1. (4) and § 94. 6. (1).

112. ܠܓܶܢܣܰܘܗ̈ܝ *according to its kinds.*

(1) ܰܘܗܝ is the form of the pronominal suffix 3rd sing. masc. with plural nouns, § 77.

(2) Rebbuy shows the plural noun. The noun singular with the 3rd masc. suffix would be ܪܽܘܚܶܗ. (See 66).

113. ܢܶܥܒܶܕ *let us make.*

(1) ܢ is the preformative of the 1st pers. plur. Imperf., § 47. Rem. 4.

(2) This Imperf. is of the *i* class ܢܶܩܛܶܠ, § 46.

114. ܐ̱ܢܳܫܳܐ—*nosho, man.* The Olaph has the linea occultans and is not pronounced. It shows the root, § 19. 1. (1).

115. ܒܨܰܠܡܰܢ *in our image.*

(1) ܒ is the inseparable preposition, § 34.

(2) ܰܢ is the pron. suffix 1st plural, § 36.

(3) ܨܰܠܡ is an *a* class segholate in the sing. masculine, §§ 67 79.

116. ܕܡܽܘܬܰܢ *our likeness.*

(1) ܰܢ as in 115. 2.

(2) ܕܡܽܘܬ is a feminine construct from ܕܡܽܘ emphatic ܕܡܽܘܬܳܐ, § 75. 8.

117. ܢܶܫܠܛܽܘܢ *let them rule.*

(1) ܢ is the sign of the 3rd plur. P^e^'al, § 45. 2.

(2) ܽܘܢ designates the masc. plur.

(3) The sign under ܠ is Mehagyono, see § 12. 1.

118. ܢܽܘܢ̈ܰܝ *fishes of.*

(1) ܰܝ is the sign of the masc. plur. constr., § 76. 3. Cf. ܐܰܒ̈ܳܗܰܝ 13.

(2) The two dots are Rebbuy.

119. ܕܪܳܚܶܫ *which creepeth.* Cf. ܩܳܛܶܠ 44. ܥܳܒܶܕ 69.

120. ܠܐܳܕܳܡ—*lo-dom, man.*

(1) Olaph throws back its vowel and quiesces, § 25. 1. (2).

(2) Lomadh is used in Syriac before the direct object, § 123.

121. ܨܰܠܡܶܗ *his image.*

(1) Absolute of noun=ܨܠܶܡ; construct, ܨܠܶܡ as in the phrase ܒܨܶܠܡ ܐܰܠܳܗܳܐ which follows: emphatic ܨܰܠܡܳܐ.

(2) ܗ is the pron. suffix "*his*" or "*of him*". Cf. ܪܽܘܚܶܗ 66.

122. ܒܪܳܝܗܝ—*b^e^roy, created he him.*

(1) ܒܪܳܐ=ܒܪܳܐ (see 2), *he created.* This is a Lomadh Olaph verb, the ܰ having been heightened to ܳ in the open syllable, § 29. 5. (1).

(2) ܝܗܝ is the pron. suffix 3rd sing. masc. with Lomadh Olaph verbs § 61.

123. ܘܟܘܒܫܘܗ̇ *and subdue it.*

(1) ܗ̇ is the 3rd fem. suffix after a form of the verb ending in a vowel § 51. E. 2.

(2) The Imv. 2nd plur. Pe'al of ܟܒܘܫ is ܟܒܘܫܘ, which before suffixes throws back the vowel from ܒ to ܟ, while the ending ܘ becomes ܘ̇.

124. ܘܫܠܘܛܘ *rule ye*, is an imperative Pe'al of the *ă* class, *i. e.* whose vowel is ̱ not ̄ as in ܟܒܘܫ. The ܘ is otiose, §§ 46, 48, 24. 2.

125. ܝܗܒܬ *I have given.*

(1) ܝܗܒ becomes ܝܗܒ before ܬ and ܬ, § 64. 7.

(2) ܬ is the preformative for the 1st sing. com., ܬ for the 3rd fem. sg.

126. ܐܝܠܢ is in the absolute state.

127. ܕܐܝܬ ܒܗ *in which are.*

(1) When the relative would take a preposition it is placed at the beginning of the clause and the preposition follows with the appropriate pronominal suffix, § 136. 6.

(2) ܐܝܬ is often indeclinable. Here the subject is ܦܐܪ̈ܐ *fruits*, §§ 65, 128.

128. ܡܐܟܘܠܬܐ *food.*

(1) Olaph is quiescent in the ̄ as is shown by the Rukhokh under the Kaph, § 10. 1. (1).

(2) ܬܐ is the fem. emphat. ending, § 76.

(3) The form is maķtul, the *a* having been obscured to *e*, § 74.

129. ܕܫܬܐ—*desh-to, which is six.*

(1) ܫܬܐ is one of the few words which have Kushoy after an unvowelled consonant in the same syllable, § 31. 1.

(2) ̄ is used as a helping vowel before a sibilant, § 33. 2, and compare § 20. *Rem.* 1.

2. Observations.

67. Most nouns take the same form before the pronominal suffixes as before the emphatic ending ܐ, *e. g.* ܪܓܠܗ, ܪܓܠܐ; ܢܦܫܗ, ܢܦܫܗܘܢ, ܢܦܫܗ̇, ܢܦܫܟܘܢ, ܢܦܫܐ.

68. The Imperf. Pe'al may have as the vowel of its 2nd radical either *a*, *i*, (which becomes *e*) or *u*, *e. g.* ܢܥܒܕ, ܢܫܟܚ, ܢܬܠ, ܢܦܩܘܕ, ܢܟܬܘܒ.

69. The Imperat. first form of the Pᵉ'al is the same as the first form of the Imperf. with the preformative omitted.

70. When new elements are added to a word, shifting of vowels frequently takes place, *e. g.* ܢܶܟܬܽܘܒ but ܢܶܟܬܒܺܝܗ, ܩܛܰܠܬ but ܩܛܰܠܬܶܗ.

71. Before the sufformatives of the Imperfect which constitute a syllable the full vowel of the 2nd radical becomes a half-vowel, *e. g.* ܬܶܟܬܒܽܘܢ.

72. Before the sufformatives ܬ and ܬ of the Perfect the vowel of the 2nd radical is dropped and the half-vowel under the first becomes ̊ *e. g.* ܩܶܛܠܶܬ.

73. Olaph may quiesce at the end of a syllabe in the middle of a word, *e. g.* ܡܐܟܘܠܬܐ. Cf. ܢܐܡܪ, ܐܡܪ, vs. 26.

74. The original fem ending was ܬ, which is retained in the construct of the noun and in the 3rd fem. sing. of the Perf., but in the fem. absolute sing. the Tau is dropped and *a* becomes *o*, *e. g.* ܡܠܟܬ ܛܠܝܬܐ, ܩܛܠܬ.

3. Grammar.

(1) Peculiarities of ܘ and ܝ, § 27.

(2) Pê Yudh and Pê Waw verbs, § 58.

(3) Review §§ 9—12, 31, 32.

4. Word Lesson.

ܝܒܫ *to be dry.*	ܝܬܒ *to sit.*
ܝܠܕ *to bear.*	ܝܢܩ *to suck.*
ܝܩܕ *to burn.*	ܥܒܕ *to make, do.*
ܝܕܥ *to know.*	ܠܐ *not.*
ܝܩܪ *to be heavy.*	ܒܪ *son.*
ܝܗܒ *to give.*	ܛܠܝܐ *child.*
ܝܪܬ *to inherit.*	ܐܢܬܬܐ *woman.*
ܡܠܟܐ *king.*	ܥܕ *until.*

5. Exercises.

1. The trees will be burnt when God shall dry the earth with His wind. 2. God said: I will make man in my image according to my likeness and I shall cause them to know what I have done. 3. Let the earth be given to man to inherit. 4. Men were not born, they were made. 5. A child has been born to us, a son has been given to us and the ruling shall be his (to him). 6. The woman sat under the tree and suckled the son whom she had borne and because he howled when he saw the sun she caused the child to sit upon the earth. 7. Thou shalt inherit the earth and thou shalt learn and know that God is very good. 8. The sun is heavier than (heavy from) the earth. 9. God will sit in the heavens and say: Let the earth and the stars be burned, let the sea be dried, and let all the lights of heaven know that I am the king who made them.

LESSON NINE. Gen. II. 1—8.

1. Notes.

130. ܘܶܫܬܰܟܠܰܠܘ *and were finished.*

(1) The vowel of the first ܘ is a helping vowel. The first syllable is half-open, §§ 33. 1, 17. 4.

(2) The last ܘ is otiose. Final ܘ does not take the linea occultans, §§ 11. 3, 24. 3.

(3) The ̈ with the 2nd radical shows the verb to be intransitive, § 41. 1.

131. ܘܫܰܠܶܡ *and (he) finished.*

(1) The vowel with the first radical shows that this is the intensive or Pa'el stem, § 41. 2.

(2) The ̈ of the 2nd radical is obscured from ̆, § 29. 2.

(3) Being without sufformative, this must be the 1st form *i. e.* the 3rd masc. sing., § 43. 4.

132. ܫܬܝܬܝܐ *the sixth.*

(1) This is an ordinal number. Cf. ܕܫܬܐ 129. See § 88. II.

(2) Ordinal numbers, like other adjectives, follow their nouns and agree in gender, number and definiteness, § 110. B.

133. ܥܒ̈ܳܕܰܘܗ݈ܝ *his works.*

(1) Rebbuy show the noun to be plural, § 13.

(2) ܘܗ݈ܝ is the 3rd sing. masc. pron. suffix, see 112, § 36.

134. ܕܰܥܒܰܕ—*da-'ebhadh, which he made.*

(1) ܕ has Kushoy after the diphthong, § 10. 1. (3).

(2) The fact that ܒ has Rukhokh shows that ܥ has a half-vowel and that the first syllable is half-open, §§ 10. 1. (2), 17. 4.

135. ܘܶܐܬܬܢܺܝܚ *and he rested himself.*

(1) This is the reflexive from ܢܳܚ to rest, §§ 41. 4, 59. 3.

(2) Since it has no sufformative, it must be the first form, *i. e.* the 3rd sing. masc., § 43. 4.

136. ܫܒܺܝܥܳܝܳܐ *the seventh.* This is the ordinal from ܫܒܰܥ *seven,* § 88. II.

137. ܩܰܕܫܶܗ—*kad-dsheh, sanctified it.*

(1) The dot over the ܕ is Kushoy and shows that the *a* is doubled. The form is intensive, the second radical being doubled, §§ 41. 2, 10. 2. (2).

(2) ܗ is the pron. suffix 3rd sing. masc. Cf. ܢܰܦܫܶܗ 66, ܕܥܒܳܕܶܗ 121.

138. ܡܶܛܽܠ *because* and ܟܽܠ *all* are the only words in which ܘ is written defectively, § 5. 5.

139. ܠܡܶܥܒܰܕ *by making.* See § 120. 1. (4) for this use of the Infinitive and compare the Hebrew.

140. ܬܰܘܠܕ̈ܳܬܳܐ *generations.*

(1) Rebbuy and ܬܳܐ designate the feminine plural emphatic.

(2) The first ܬ is prefixed, the form of the noun being ܬܰܘܠܶܕ. The root is ܝܠܶܕ *to bear,* § 74. 3.

141. ܘܰܕܐܰܪܥܳܐ *and of the earth.*

(1) The inseparable particles retain their helping vowel, when the succeeding consonant has a vowel thrown back from a following Olaph; so *wa-dhar-'o,* not *wdhar-'o,* § 34. 3. *Rem.* 2.

(2) When a noun is in connection with two or more succeeding genitives, each of the latter is preceded by ܕ, § 97. A. *Rem.* 2.

142. ܐܶܬܒܪܺܝܘ *they were created.*

(1) ܐܬ is the sign of a reflexive or passive stem. The absence of a vowel after ܬ or ܦ show it to belong to the simple passive or Ethpe'el, § 41. 4.

(2) Waw designates the 3rd plural, § 43.

(3) Yudh shows that the root is Lomadh Yudh (called Lomadh Olaph, see § 60).

(4) For the diphthong *iu*, see § 8. 1. (3).

143. ܕܰܥܒܰܕ *in which (he) made.* The preposition is often omitted from the relative clause, § 136. 6. *Rem.* 1.

144. ܗܘܰܘ *(they) had been* is in the 3rd plur. Perfect from ܗܘܳܐ fem ܗܘܳܬ. The two dots under the word are meant to distinguish the verb as denoting *existence* from the same verb used as an enclitic copula. Compare: ܗܘܐ vs. 6 with ܗܘܐ vs. 7 and ܗܘܘ vs. 25.

145. ܐܺܝܬܳܐ *had gone out.*

(1) The first Olaph is prosthetic, § 20. 1.

(2) Yudh quiesces in ̈ according to §§ 25, 3, 58. 1. *Rem.* 2.

(3) The form is the 3rd masc. sing. Pe'al, the verb being both Pê Yudh and Lomadh Olaph, §§ 58, 60.

146. ܐܰܚܶܬ *(he) caused to come down.*

(1) The form is Aph'el from ܢܚܶܬ, the Nun being assimilated. Cf. ܐܰܦܶܩ 69. It is the first form of the Aph'el *i. e.* 3rd masc. sing. Perfect, § 53. 2.

(2) ̈ is derived from an original ̈, § 29. 2.

147. ܠܰܝܬ *was not,* is contracted from ܠܳܐ & ܐܺܝܬ, § 65. Tau has Kushoy after the diphthong, § 10. 2. (3).

148. ܡܰܒܽܘܥܳܐ *mist.* The root is ܢܒܰܥ *to well.* The form is ܡܰܢܒܽܘܥ, the Nun being assimilated and the ܒ doubled. Cf. ܡܰܕܢܚܳܐ 128 and see §§ 18. 1, 74. 2. (5).

149. ܣܳܠܶܩ ܗܘܳܐ *used to go up.*

(1) ܣܳܠܶܩ is the active part. Pe'al first form; see 46. It is in the absolute state because a predicate, § 93. 3. (2) *a.*

(2) ܗܘܳܐ is enclitic and hence the ܗ has the linea occultans and is unpronounced §§ 64. 5, 127. 1. Cf. ܗܘܰܘ 144, ܗܘܳܬ, ܘܗܘܳܐ vs. 19.

M

(3) The Perf. of ܗܘܳܐ after the part. denotes continuous or repeated action, or state, § 127. 3. (3).

150. ܡܰܫܩܶܐ ܗܘܳܐ *was watering.*

(1) The construction is the same as that in 149. 2, 3.

(2) The Part. is the first form of the Aph'el, as is shown by ܡ prefixed with ܰ, §§ 41. 3, 50. 2.

(3) The first form is ܫܩܳܐ. Cf. ܗܘܳܐ, ܚܙܳܐ.

151. ܒܰܐܦ̈ܰܘܗ݈ܝ *in his nostrils.*

(1) ܒ is the preposition with the vowel of Olaph drawn back.

(2) Olaph quiesces according to § 25. 1. (2).

(3) ܰܘܗ݈ܝ is the ending of the 3rd masc. sing. suffix with plural nouns. § 77.

(4) The two dots over ܘ are Rebbuy; the one over ܒ is Kushoy after a consonant; the one over ܦ is Kushoy denoting the doubling to compensate for the assimilated Nun, §§ 10, 12.

152. ܚ̈ܰܝܶܐ *life,* lit. *lives.*

153. ܡܶܢ ܩܽܘܕܡܳܐ *from (the) front, i. e. from the east.*

154. ܣܳܡ *(he) put.* This is the first form of a verb Ê Wau contracted from ܣܘܰܡ, § 59.

2. Observations.

75. Most verbs have ܰ in the first form; some intransitive verbs have ܶ; two verbs have ܘܽ, § 41. 1.

76. The vowel occurs everywhere in the Pa'el stem after the first radical.

77. Ordinal numbers are formed from the radicals of the cardinals by putting a half-vowel after the first radical, ܳܝ after the second, and ܳܝܳܐ after the third, § 88.

78. The inseparable particles take a helping vowel before an unvowelled consonant and form with it a half-open syllable.

79. Whether a form is Pᵉ'al or Pa'el depends often upon the *usus loquendi, e.g.* ܩܰܛܠܶܗ may be either ḳatleh, or ḳattᵉleh, *i. e.* simple or intensive. The sense and not the writing (which is the same in both) determines the stem.

80. Nouns may be formed by prefixing ܡ or ܬ to the root, *e. g.* ܡܰܦܩܳܢܳܐ, ܬܰܫܒܽܘܚܬܳܐ.

81. The Imperative has only a 2nd person.

3. Grammar.

(1) The Perfect of the verb with suffixes, § 51. A. B.

(2) Quantity of vowels, § 28.

(3) Review, §§ 43, 44.

4. Word Lists.

ܫܒܩ *to forsake.*
ܩܛܠ *to kill.*
ܫܩܠ *to weigh.*
ܡܠܟ *to counsel.*
ܐܚܕ *to seize.*
ܪܕܦ *to follow.*
ܬܡܗ *to wonder.*
ܒܛܢ *to conceive.*
ܫܒܚ *to glorify.*
ܦܩܕ *to command.*

5. Exercises.

1. God has forsaken you because you forsook him. 2. God weighed his works which he had done and when He saw that he had not followed Him He killed him. 3. I counsel thee to seize them (ܕ with the Imperfect), because they have not glorified thee. 4. He made me to wonder (Aphel) when he commanded us to kill them because they had followed thee. 5. She conceived me and bare me. 6. I followed him and seized him and killed him because thou didst command me. 7. God rested from all his works when He had completed them and He blessed them and sanctified them. 8. When the heavens and the earth were created, a tree did not exist in the earth. 9. The Lord caused rain to come down and the herb of the field sprang up, and a mist was going up to water all the trees of the field and every green herb (greenness of herb) which Adam had for food, (which to Adam were).

LESSON TEN. Gen. II. 9—15.

1. Notes.

155. ܘܐܘܥܝ *and he caused to go out.*

(1) The Olaph designates the Aph'el stem, § 41. 3.

(2) The first radical is ܘ. This becomes Yudh in the simple and intensive stems, § 58. The ܝ comes from *iy*.

(3) The third radical was originally Olaph; but most verbs of this kind have gone over into regular Lomadh Olaph verbs, § 60.

156. ܕܐܪܓܝܓ *which was pleasant.*

(1) This is a relative clause, § 136.

(2) The Olaph is prosthetic, § 20. 1.

(3) The noun is of the passive participial form ܦܥܝܠ. Being a predicate it is in the absolute state, § 93. 3. (2).

157. ܠܡܚܙܐ *to see.* This is the Infin. const. Pᵉ'al of ܚܙܐ; the ̇ of ܡܦܥܠ becoming heightened in the opened syllable.

158. ܠܡܐܟܠ *to eat.* Pê Olaph verbs form their Infinitives regularly except that the Olaph quiesces and the following consonant is consequently aspirated. Cf. 128.

159. ܒܡܨܥܬܗ ܕܦܪܕܝܣܐ *in the midst of the garden.*

(1) For the genitive construction compare ܪܘܚܗ ܕܐܠܗܐ 15. See § 97 B.

(2) For ܡܨܥܬ and ܡܨܥܬܗ, see 42 and 15.

160. ܠܡܫܩܝܘܬܗ *to water it.*

(1) ܠ is the preposition; ܗ the pron. suffix 3rd sing. masc.

(2) ܡ is the sign of the Part. and Infin. of the Aph'el, §§ 49, 50.

(3) ܝܘܬ designates the Infin. construct; the absolute would end in ܝܘ. Cf. ܡܥܡܘܕܘ 85.

(4) Yudh belongs to the root, which is Lomadh Yudh (Olaph), § 60.

161. ܠܦܪܕܝܣܐ *to wit, the park.* This is in apposition with ܗ; the ܠ may in such cases be rendered by "*to wit*", see § 123. 2. (7).

162. ܗܘܐ ܠ *becometh.* The verb ܗܘܐ followed by ܠ may be translated by *become.* This is the Pᵉ'al Part. See § 60. 5.

163. ܐܪܒܥܐ ܪ̈ܫܝܢ *four heads.*

(1) The cardinal generally precedes.

(2) The noun following may be in either the emphatic or absolute state, § 110, A. 1.

164. ܫܡܗ ܕܚܕ *the name of it which is one.*

(1) On ܫܡ, see § 87. 29.

(2) ܕܚܕ is a relative phrase limiting the pronominal suffix and not the noun, see § 136. 4.

(3) This clause takes the place of the ordinal, § 110. B.

165. ܕܡܚܕܪ ܗܘ *It is that, which is surrounding.*

(1) ܕ = that which, see § 104. 2. (2) *Rem.*

(2) The phrase is a predicative substantive clause, § 135. 2.

(3) The Participle has ̱ instead of ̈ because of the ܪ, see § 52. 3.

(4) ܗܘ is the demonstrative pronoun 3rd masc. sing. § 35. It has the point over to distinguish it from ܗܘ, § 6. 6. (1).

166. ܠܟܠܗ *all.*

(1) ܠ is the sign of the direct object, § 123.

(2) The participle governs a noun.

167. ܕܬܡܢ *where,* introduces a relative clause, §§ 104. 4, 136.

168. ܘܕܗܒܗ—*w^e^dha-h^e^bhoh, and the gold of it.* The Rukkokh under the ܒ shows that the noun is not a segholate, but one which had originally two short vowels, *i. e. dahabh,* § 68. The segholate would be ܕܗܒܗ *dah-boh,* like ܓܢܣܗ *gen-so.*

169. (1) ܗܝ *that,* is a demonstrative pronoun limiting land. It follows its noun when attributive and agrees with it in gender and number.

(2) The point above the Hê stands for *o* and shows that *hoy* not *hî* is to be read, § 6. 6. (2) *b.* Compare Gen. III. 12 for ܗܝ.

170. ܛܒ *good,* is the predicative adjective. It agrees with its subject in gender and number, but not in definiteness, § 93. 3. (2).

171. ܬܢܝܢܐ *the second,* is an ordinal form for ܬܪܝܢ. Above in vs. 11 and below in vs. 14, the cardinal preceded by ܕ is used in its stead, § 110. B.

172. ܠܐܕܡ *the man,* ܠ with the direct object, § 123.

173. ܫܒܩܗ—*shabh-ḳeh, left him.*

(1) The Rukkokh under the ܒ shows that it is not doubled. The form is, therefore, not Pa'el, but P^e^'al. Cf. 137 and see, § 41. 2.

(2) ܗ̈ is the pron. suffix. 3rd sing. masc.

174. ܕܢܦܠܚܝܘܗܝ *that he might till it.*

(1) ܕ introduces the clause of purpose, § 137. 4.

(2) ܝܘܗܝ is the form of the 3rd sing. pron. suffix with the Imperf., § 51. D. 2.

(3) In ܢܶܦܠܽܘܚ (from ܢܶܦܠܽܘܚ § 46) Nun is the sign of the 3rd pers.; the absence of sufformatives shows it to be masc. sing.; the ̂ with the preformative shows the simple stem, §§ 45. *Rem.* 2, 47. *Rem.* 4.

(4) The 1st plur. Imperfect would also be ܢܶܦܠܽܘܚ. The context alone can determine whether the 1st or 3rd person is meant, § 37. *Rem.* 5.

175. ܘܢܶܛܪܺܝܗ̇ *and keep it.* This is the same in every respect as the preceding, except that we have ܢܶܛܰܪ for ܢܶܢܛܰܪ the radical Nun being assimilated. Cf. ܐܶܦܶܩ 72, and see § 53. 2.

2. Observations.

82. Lomadh Olaph verbs are mostly those which were originally Lomadh Wau or Yudh.

83. The conjunction Wau, the inseparable prepositions ܒ and ܠ and the relative ܕ, take ̗ before a consonant with a half-vowel.

84. The vowel under the 2nd radical of the Peʿal Imperfect is dropped before suffixes and before sufformatives forming a new syllable.

85. A short vowel may be dropped, volatilized or shifted.

86. A naturally long vowel is unchangeable, § 73.

3. Grammar.

(1) Euphony of vowels, § 29.

(2) The Imperfect &c. of the regular verb with suffixes, § 51. C. D. E. F.

(3) Review §§ 36 and 45—47.

4. Word Lesson.

ܐܶܢ *if.*

ܫܡܰܥ *to hear.*

ܐܶܫܬܰܡܰܥ *to obey.*

ܒܰܪܶܟ *to bless.*

ܪܕܰܦ *to follow.*

ܩܰܕܶܫ *to sanctify.*

ܦܠܰܚ *to till.*

ܥܒܰܕ *to serve.*

ܓܢܰܒ *to steal.*

ܫܰܠܶܡ *to deliver.*

ܒܥܶܠܕܒܳܒܳܐ *enemy.*

5. Exercises.

1. God will bless you if ye will follow him and serve him. 2. I will cause it (fem.) to bring forth herbs and trees which are pleasant (pl.) to see and whose fruits (which their fruits) are good for eating. 3. The river shall water it (masc.). 4. Let him bring and leave him in Eden that he may till it and keep it. 5. I shall cause it to surround all the land of Cush. 6. Thou wilt bless us and wilt cause us to see God. 7. I shall kill you; thou wilt bless him; she will forsake you. 8. Thou (fem.) wilt kill him if he shall not obey me. 9. Let him hear me, God said, and I shall bless him and I will sanctify him. 10. Thou (fem.) wilt hear me and obey me and I will hear thee when thou callest. 11. They will steal him and will kill him and will deliver him to his enemies.

LESSON ELEVEN. Gen. II. 16—20.

1. Notes.

176. ܘܶܐܡܰܪ ܠܶܗ *and said to him.* For the indirect object, see § 124. For the form, see § 34. 2.

177. ܡܶܐܟܰܠ ܬܶܐܟܽܘܠ *thou mayest eat.*

(1) ܡܶܐܟܰܠ is the absolute Infinitive used to intensify the idea of the verb, § 119. 1. The Rukkokh under the Kaph shows that Olaph is quiescent, § 10. 1.

(2) The verb is in the 2nd masc. sing. like ܬܶܩܛܽܘܠ except that the Olaph is quiescent, § 55. 2.

178. ܠܳܐ ܬܶܐܟܽܘܠ *thou shalt not eat.* This is the negative of the preceding. The negative of the Imperative is expressed by the Imperfect preceded by ܠܳܐ, §§ 114. 1. (2), 115. 3.

179. ܡܶܛܽܠ ܕ *because that.*

(1) 'Eṣoṣo is written defectively, § 6. 5.

(2) This is a common way of introducing the causal adverbial clause, § 137. 5. (2).

180. ܕܬܶܐܟܽܘܠ *in which thou eatest.* The preposition with its pro-

nominal suffix is omitted, as frequently in temporal clauses, § 136. 6. *Rem.* 1.

181. ܬܡܘܬ ܡܵܘܬܵܐ *the death shalt thou die.*

(1) ܡܵܘܬܵܐ is a segholate noun of the *a* class; here used instead of the Infinitive absolute, § 119. 2. *Rem.* 1.

(2) ܬܡܘܬ is the 2nd person masc. sing. Imperf. Pᵉʿal for ܬܶܡܘܽܬ, *wu* going over into ܘܿ and the helping vowel of the preformative being volatilized. The root is Ê Wau, §§ 29. 7. (1), 59. 2.

182. ܕܢܶܗܘܶܐ *that should be &c.*, is a substantive subject clause, to which ܠܐ ܫܰܦܺܝܪ is the predicate, § 135. 1, 2.

183. ܒܰܠܚܘܽܕܰܘܗܝ *alone.* This is a compound of the prepositions ܒ and ܠ with the noun ܚܘܽܕ *unique*, followed by the pronominal suffix, § 89. B. *Rem.* 3. 3.

184. ܐܶܥܒܶܕ *I will make.*

(1) The Olaph is the preformative for the first person sing. Imperf.

(2) The ܶ under the second radical signifies an Imperf. of the *i* class, § 46. 1 and cf. יִתֵּן and יֵשֵׁב in Hebrew.

185. ܡܥܰܕܪܳܢܳܐ *help.*

(1) The point over the ܕ shows that the form is intensive Dolath being doubled, § 41. 2.

(2) The ܡ points to a participial form of the Paʿel stem, § 50. 2.

(3) The ending ܳܢܳܐ is often appended to participles to make *nomina agentis*, § 75. 1.

186. ܐܰܟ݂ܘܳܬܶܗ *like him.* This is the form which ܐܰܝܟ takes before suffixes, § 89. B. *Rem.* 3. 1.

187. ܘܰܐܝܬܺܝ *and he brought.*

(1) The first form is ܐܶܬܳܐ, § 64. 4, a Pê Olaph and Lomadh Olaph verb.

(2) The Olaph of ܐܝܬܺܝ denotes the Aphʿel stem, § 41. 3.

(3) For the ending ܺܝ see 155.

188. ܕܢܶܚܙܶܐ *that he might see.*

(1) The ending ܶܐ (like ֶה in Hebrew) is the common ending for Lomadh Olaph Imperfects, § 60. 3.

(2) The clause denotes purpose, § 137. 4. Cf. 174.

189. ܡܳܢܳܐ ܩܳܪܶܐ *what he was calling.*

(1) This is an indirect question introduced by the interrogative pronoun ܡܳܢܳܐ, § 132. 6. *Rem.* The sentence is an object substantive clause, § 135. 3. (2).

(2) On ܩܳܪܶܐ see 162 and § 60. 5.

190. ܗܰܘ ܗܽܘ—*hau hu, that is.*

(1) ܗܰܘ with a dot above the ܗ is *hau* the demonstrative pronoun, ܗܽܘ with a dot below the ܗ is the personal pronoun *hu*, §§ 6. 6. (1), 35, 37.

(2) The demonstrative ܗܰܘ resumes and is in apposition with the substantive clause, beginning with ܟܽܠ, which precedes it. The clause with ܟܽܠ is equivalent to a noun absolute, §§ 95. 3, 135. 1.

(3) ܗܽܘ is the copula, § 101.

191. ܫܡܳܗܶܐ *names.* This is an irregular plural from ܫܡܳܐ *name,* §§ 86. 14, 87. 29.

192. ܠܳܐ ܐܶܫܬܟܰܚ *there was not found.*

(1) ܫ and ܬ have been transposed, § 21. 1.

(2) It is Ethpe'el as is shown (*a*) by the absence of a vowel before or after the ܬ (which her after transposition takes the place of the first radical), (*b*) by the non-doubling of the second radical.

(3) By there being but two syllables; the intensive and causative passive having three.

(4) According to form, this might be the 1st pers. sing. Imperf., or the 3rd masc. Perf.: the *sense* requires the latter.

2. Observations.

87. Imperfects may have *a*, *i*, or *u*, under the 2nd radical of the Pe'al.

88. The same form is often used in different senses, the sense in a particular case is to be determined by the context, *e.g.* ܐܶܫܬܟܰܚ, ܬܶܫܬܟܚܽܘܢ.

89. Clauses are substantive, adjective, or adverbial, § 135.

90. Notice the difference in mood denoted by the Imperfect in vs. 16—18, § 114.

3. Grammar.

(1) Lomadh Olaph verbs, § 60.

(2) Lomadh Olaph verbs with suffixes, § 61.

(3) Review, § 27.

4. Word Lesson.

ܚܙܐ *to see.*
ܣܛܐ *to decline.*
ܚܕܝ *to rejoice.*
ܢܚ *to be at rest.*
ܢܣܝ *to tempt.*
ܒܐܫ ܠ *It displeased.*
ܐܝܟ *like* (before suffixes ܐܟܘܬ, § 89. B. 1).
ܫܕܐ *to cast.*
ܨܒܐ *to will, wish.*
ܩܪܐ *to call.*
ܡܠܐ *to be full.*
ܚܘܝ *to show.*
ܨܠܝ *to pray.*
ܐܘܪܚܐ *way.*
ܕܟܝ *to purify.*
ܐܘܪܒ *to magnify.*
ܨܒܝܢܐ *will.*
ܢܦܫܐ *soul.*
ܩܘܫܬܐ *truth.*
ܚܕܘܬܐ *joy.*

5. Exercises.

1. Adam was commanded to call names to all the beasts. 2. He rejoiced when he saw that God had created the woman (for a) helper corresponding to him. 3. Ye have declined from the way and have tempted God who wished to give you rest (V stem). 4. It displeased God that Adam hid himself in a tree which was in Paradise, because he had eaten of the tree of the knowledge of good and of evil. 5. Let us pray to God that he may show to us his will and that he may not cast us from paradise. 6. Rejoice, my soul, and magnify the Lord God because he hath heard thee when thou didst pray to him. 7. Purify thy way and decline not from the truth, rejoice and cause thy soul to rest in God and he will fill thee (with) joy and show thee his truth and thy soul shall be purified. 8. When God shall see that it is not good

that I shall be alone he will make for me a helper corresponding to me. 9. God formed them and brought them to Adam that He might see what he was calling them.

LESSON TWELVE. Gen. II. 21—25.

1. Notes.

193. ܘܐܪܡܝ *and he cast.* Aph'el Perf. 1st form. Cf. ܐܘܒܕ 155, ܐܝܬܝ 187.

194. ܘܕܡܟ *and he slept.*

(1) Wau has a helping vowel and with it forms a half-open syllable, §§ 17. 4, 33. 2.

(2) ܕܡܟ instead of ܕܡܟ because intransitive, § 41. 1. (2).

195. ܘܐܘܚܡ *and he closed.* Aph'el 1st form. Cf. 193, 155, 187.

196. ܚܠܦܝܗ̇ *in place of it.* ܚܠܦ like many other prepositions takes the plural construct form before the pron. suffixes, § 77. 4.

197. ܕܢܣܒ *which he had taken.*

(1) The clause is adjective, § 136.

(2) The Perfect is used in the sense of our Pluperfect, § 112. 1. (3).

198. ܠܐܢ݇ܬܬܐ *to a woman.*

(1) The preposition ܠ draws back the vowel the Olaph quiescing, § 34. 2.

(2) The line with the Nun is linea occultans, § 11.

(3) The word is the indirect object, the verb governing two objects, § 125. 3.

199. ܘܐܝܬܝܗ̇—*way-tᵉyoh, and he brought her.*

(1) ܗ̇ is the pron. suffix 3rd fem. (Cf. ܢܦܫܗ̇ 102), § 61.

(2) ܐܝܬܝ is the same form as ܐܝܬܝ 190, the original consonantal *y*, remaining before the suffix the preceding vowel having been volatilized *i. e. aytî* becomes *ay-tᵉyoh*, § 7. 3. (2) *b*.

200. ܗܢܐ ܙܒܢܐ *this time.*

(1) ܗܢܐ is a demonstrative pronoun, § 37. 1.

(2) The pronoun may precede or follow its noun, § 102. 1.

201. ܓܪܡܝ *my bones.* The form of the const. plur. masc. is the same as that of the const. plur. masc. with the suffix 1st sing., § 77. 1.

202. ܒܣܪܝ—*besrᵉ, my flesh.*

(1) An appended Yudh designates *my,* § 36.

(2) The final Yudh in words like this is pronounced like *e,* § 31. 3, *Rem.* 1.

203. ܢܣܝܒܐ (*was she*) *taken,* is the passive Part. Pᵉʿal fem. sing. absolute, § 50. 1, § 76. 2.

204. ܢܫܒܘܩ (*he*) *shall forsake.*

(1) The Nun prefixed denotes the 3rd person of the Imperfect.

(2) The ܘ shows it is an Imperfect in *u* of the simple stem, § 46. 3. *Note.*

205. ܠܐܒܘܗܝ—*la-bhu, his father.*

(1) ܗܝ is the pronominal suffix 3rd sing. masc. after a vowel, § 36.

(2) ܐܒܘ is the form of ܐܒܐ before suffixes, § 87. 1.

(3) The Olaph throws back its vowel to the Lomadh and quiesces in the Pᵉthoḥo, §§ 32. 3, 25. 2.

(4) Lomadh is used in Syriac before the direct as well as before the indirect object, § 123. 2. (3).

206. ܘܢܩܦ *and he shall cleave.*

(1) This is the 3rd. pers. sing. masc. Imperf. form ܢܩܦ, the Nun having been assimilated. It is to be pronounced *nekkaph,* from *nenkaph,* §§ 18, 53.

(2) Notice that there is no Waw conversive in Syriac.

207. ܬܪܝܗܘܢ *the two of them.* ܬܪܝ is the construct of the Dual ܬܪܝܢ 52. On the Dual in Syriac, see § 76. 5.

208. ܚܕ ܒܣܪ *one flesh.*

(1) For the order see, § 99. 1. *Rem.* 1.

(2) The predicate noun adjective or participle is commonly put in the absolute state, § 93. 4. (2).

209. ܥܪܛܠܝܢ *naked.*

(1) The ܝܢ is the sign of the masc. plur. absolute, §§ 76. 3, 93. 4. (2).

(2) Notice the coïncidence of the diacritical point of the *r* with one point of Rebbuy, § 13. 2.

210. ܒܗܬܝܢ—*boh-tîn, ashamed.*

(1) This is the Act. part. of the simple stem in the abs. plur., §§ 50. 1, 76. 3.

2) The singular is ܩܗܬ, but the short vowel *e* is lost and the Tau hardened when an affix is appended. See § 30. 1, and compare § 31. 3. *Rem.* 1.

2. Observations.

91. There is no Waw conversive or consecutive in Syriac.

92. Syllables may be open, closed, or half-open, § 17.

93. *Aw* does not contract into *ô* in Syriac.

94. A Dual occurs in a few instances.

95. Changeable vowels may be dropped in inflection, § 7. 3. *e. g.* ܩܗܬܝܢ, ܐܠܦܗ.

96. The predicate adjective agrees with its antecedent in gender and number, but not in state, *e. g.* ܢܦܝܫܐ, ܕܢܗܝܪܝܢ.

97. The rules for the assimilation of Nun are the same in Syriac as in Hebrew.

98. Lomadh may be used in Syriac before the direct as well as before the indirect object.

99. Attributives usually follow the nouns, but occasionally they precede.

100. Attributives agree with their nouns in gender, number, and state.

101. In stative verbs, the vowel in usually *e.*

102. Some prepositions take the plural form before suffixes.

3. Grammar.

(1) Pê Olaph Verbs, § 55.

(2) Peculiarities of Gutturals, § 26.

(3) Review, §§ 24, 25, 76, 77.

4. Word Lesson.

ܐܚܕ *to seize.* V. *to close.*

ܐܙܠ *to go out.*

ܐܬܐ *to come.*

ܐܝܬܝ *to bring.*

ܝܠܦ *to learn.* III. *to teach.*

ܐܒܠ *to mourn.*

ܐܣܪ *to bind.*

ܥܠ *upon, for* (before suffixes ܥܠܝ. See § 77. 4.).

ܕܟܪ *to remember.*

ܩܕܡ *before* (Plural form before suffixes).

5. Exercises.

1. Adam slept because a sleep had been cast upon him and one of his ribs was taken and the flesh was closed in place of it; and the rib which had been taken from Adam was formed into a woman whom God brought to Adam. 2. And when Adam saw her he said: this shall be called woman and because that she is flesh of my flesh shall the two of us be one flesh. 3. Go ye out and say to the woman that I will bind her and teach her not to mourn for her. 4. Bring her to me and I will teach her to go out and to come in before thee. 5. Remember God and he has remembered thy father and thy mother. 6. They shall be blessed who mourn. 7. His flesh was eaten. 8. I shall learn all that thou wilt teach me because I am not ashamed to learn. 9. I shall teach and ye shall learn all that is written in this good book.

LESSON THIRTEEN. Gen. III. 1—5.

1. Notes.

211. ܥܪܝܡ ܗܘܐ *was cunning.*

(1) ܗܘܐ is enclitic after a participial adjective predicate, § 127. 10, and hence the ܗ has the linea occultans, § 11.

(2) The participle is the simple passive in the absolute singular, § 50. 1.

212. ܡܢ *from.*

(1) The point beneath shows that it is to be read *men* not *man* or *mon*, § 6. 6. (1).

(2) *Men* after the adjective denotes the comparative, § 101. 1.

213. ܟܠܗ *every.*

(1) The point above shows that the suffix is the feminine ܗ and not the masculine ܗ. See 102 and § 6. 6. (2) *b*.

(2) For the construction, see §§ 108. 1. (4), 97. B. *Rem.* 4.

214. ܕܥܒܰܕ *which (he) had made.*

(1) ܕ introduces the relative or adjective clause which limits ܚܰܝܘܬܳܐ, § 136.

(2) The Perfect here denotes our Pluperfect, § 112. 1. (3).

215. ܚܶܘܝܳܐ is in apposition with ܥܰܪܺܝܡܳܐ, § 94. 1

216. ܫܰܪܺܝܪܳܐܺܝܬ *truly.*

(1) ܐܺܝܬ is the common ending for adverbs, § 89. A. 3.

(2) ܫܰܪܺܝܪ *sharrîr* is of the formative ḳaṭṭîl, § 72. 2. (4).

217. ܐܶܡܰܪ *hath (he) said.* The Perfect is the Present Perfect, § 112. 1. (2).

218. ܕܠܳܐ ܬܶܐܟܠܽܘܢ *that ye shall not eat.*

(1) ܕ introduces the objective substantive sentence, which is here a quotation, § 135. 3. (3).

(2) ܠܳܐ with the Imperfect may be either "ye shall not" or "eat not". § 114. 1.

219. ܠܚܶܘܝܳܐ *to the serpent.* The indirect object is introduced by Lomadh, § 124.

220. ܕܡܶܢ *from.* ܕ introduces the quotation like *ut* in Latin, § 135. 3. (4).

221. ܕܰܒܦܰܪܕܰܝܣܳܐ *which are in the Paradise.*

(1) ܕ introduces the relative or adjective clause, § 136 and is the subject of the nominal sentence, § 130.

(2) The copula is supplied and "in the Paradise" is the predicate, § 130.

222. ܟܽܠܗܽܘܢ *all of them,* is a clause in apposition with ܐܺܝܠܳܢܶܐ, § 94. 1.

223. ܢܶܐܟܽܘܠ *we may eat.* For the use of the Imperfect as our Potential mood, see § 114. 2.

224. ܕܠܳܐ ܬܡܽܘܬܽܘܢ *lest ye die.* This is an adverbial clause of result, § 137. 4.

225. ܡܡܳܬ is an Inf. Absolute from ܡܺܝܬ *to die.* It is here used adverbially to strengthen the cognate verb following, § 119. 1. (1) *a.* Compare the Hebrew.

226. ܡܶܛܽܠ ܕ *because that,* introduces an adverbial clause of cause or reason, § 137. 5. (2).

227. ܝܕܥ (*he*) *knows* is the Perfect used as a Present like the Greek 2nd Perfect, § 112. 2. (1).

228. ܕܒܝܘܡܐ ܕ *that in the day that.*

(1) The ܕ introduces an objective substantive clause, § 135. 3.

(2) ܒܝܘܡܐ ܕ introduces an adverbial clause of time, § 137. 2. The whole clause is equivalent to "when" and introduces the protasis, § 138. 3. (3).

229. ܕܐܟܠܝܢ *in which* [*ye*] *shall be eating of.*

(1) The Participle here denotes a state or continuous action. The time is made future by the clause "in the day in which", § 116. 1. (2).

(2) The ܕ is a relative adjective agreeing with ܝܘܡܐ, § 104. 1. (4).

230. ܡܬܦܬܚܢ—*mith-pat-tᵉḥon, shall be opened.*

(1) Participle formed by ܡ prefixed, § 50. 2; the first ܬ shows the Reflexive, § 41. 4; the Kushoy over the second ܬ shows the Intensive, § 41. 4; the ܢ shows the fem. plur., § 76. 4.

(2) The Participle is in the future tense since it introduces an apodosis, which depends on a protasis which is not yet fulfilled.

231. ܥܝܢܝܟܘܢ *your eyes.* The noun is in the dual construct, § 76. 5. (2). Since the participle has no dual, it is put in the plural, § 99. 2.

232. ܗܘܝܬܘܢ *ye shall be.* This is the Act. Pᵉ'al Part. plural and the 2nd pers. plur. personal pronoun, which have coalesced. See § 35. 2. *Note.*

233. ܝܕܥܝ *knowers of.* The Part. Act. Pᵉ'al in the construct plural before an object, § 118. 2. The order of time is not involved in the form but only in the connection, § 116.

2. Observations.

103. ܗܘܐ when it follows the predicate is enclitic and the Hê is unpronounced securing the linea occultans, vs. 1; but when the subject precedes or the sentence is verbal the Hê is pronounced, see Gen. 1. 2.

104. The predicate is in the absolute state, unless it be a noun which has no absolute state or unless it is meant to be specially determined.

104 *a*. The comparative is usually expressed by putting the adjective first, in agreement as to gender and number with the noun to be compared, *e. g.* ܚܰܟܺܝܡ is the adjective, and ܚܶܘܝܳܐ is the noun to be compared. The idea with which the comparison is made is preceded by ܡܶܢ, *e. g.* ܟܽܠ ܚܰܝܘܳܬܳܐ.

105. Observe that a point above denotes *ă* or *o* as distinguished from *e*, *e. g.* ܡܶܢ=*men*, but ܡܰܢ=*man* or *mon*, ܟܽܠܳܗ=*kulloh*, ܟܽܠܶܗ=*kulleh*.

106. ܕ may be either a demonstrative pronoun like זֶה, or a relative pronoun (like זוּ or זֶה used relatively) or a conjunction.

107. Observe that there are in this lesson three ways of expressing the genitive relation. (1) ܚܰܝܘܳܬܳܐ ܕܕܰܒܪܳܐ vs. 1. (2) ܦܺܐܪܰܝ ܐܺܝܠܳܢܶܐ vs. 2. (3) ܡܶܨܥܬܶܗ ܕܦܰܪܕܰܝܣܳܐ vs. 3.

108. The form ܩܛܰܠ in this lesson denotes, (1) a simple past, *e. g.* ܗܘܳܐ vs. 1, (2) a pluperfect, ܥܒܰܕ vs. 1, (3) a present perfect, *e. g.* the second ܐܶܡܰܪ in vs. 1, (4) a present (Greek 2nd perfect) ܝܳܕܰܥ vs. 5.

109. We have in this lesson specimens of the three kinds of dependent sentences, (1) substantive *e. g.* in the object clauses beginning with ܕܠܳܐ vs. 1, ܕܡܶܢ vs. 2, the first ܕܠܳܐ vs. 3, ܕܒܝܰܘܡܳܐ vs. 5, (2) adjective *e. g.* in the relative clauses beginning with ܕܥܒܰܕ vs. 1, ܕܒܦܰܪܕܰܝܣܳܐ vs. 2, ܕܒܡܶܨܥܬܶܗ vs. 3, ܕܐܳܟܠܺܝܢ vs. 5, (3) adverbial, *e. g.* in the clause of result ܕܠܳܐ ܬܡܽܘܬܽܘܢ vs. 3, and in the causal clause ܡܶܛܽܠ ܕ vs. 5.

110. There is no way in Syriac to distinguish between the negative of the Imperfect and that of the Imperative, *i. e.* ܠܳܐ ܬܶܩܛܽܘܠ is "*thou shalt not kill*" or "*kill not*". ܠܳܐ is both οὐ and μή, לא and אל.

111. All the modes may be expressed by the Imperfect. In this lesson we have ܠܳܐ ܬܺܐܟܠܽܘܢ ye *shall* not eat, vs. 1, ܢܺܐܟܽܘܠ we *may* eat, vs. 2, ܠܳܐ ܬܡܽܘܬܽܘܢ lest ye *die*, vs. 3. ye *shall* not die vs. 4.

112. Words denoting members of the body which occur in pairs and a few other words have a separate form for the dual in the absolute state. In all other cases the dual has disappeared, the plural taking its place.

3. Grammar.

(1) Ê Wau verbs, § 59.

(2) Peculiarities of Wau, § 27.

(3) Review, §§ 27, 29, 58, 60.

4. Word Lesson.

ܩܳܡ *to stand, arise.*

ܣܳܡ *to put.*

ܡܺܝܬ *to die.*

ܠܳܛ *to curse.*

ܢܳܕ *to shake.*

ܙܳܥ *to be moved.*

ܥܳܪ *to watch.*

ܥܳܩ II. *to be anxious.*

ܛܰܝܶܒ *to prepare.*

ܛܰܢܶܦ *to defile.*

ܕܝܳܪܳܐ *habitation, dwelling.*

ܐܺܝܕܳܐ *hand*, § 87. 2.

ܪܺܫ *head.*

ܩܝܳܡܳܐ *covenant.*

ܕܠܳܐ *lest.*

ܡܕܺܝܢܬܳܐ *city.*

ܩܕܳܡ ܕ *before.*

5. Exercises.

1. The beast of the field ate the fruits of the trees which (were) in the midst of the paradise. 2. The woman said to the serpent that the beast which (was) in the field prudent was from [was more prudent than] every serpent of paradise. 3. The man shall eat of the fruits which (are) in the field all of them. 4. I shall not eat of it because I know that in the day that I eat (§ 35. 2) of it I shall surely (abs. Inf.) die. 5. Ye have cursed God and he will prepare a habitation for you. 6. The woman stood and put her hand upon my head and said: Accursed (be) thou and mayest thou die (thou shalt die) because thou hast been defiled. 7. Watch ye (II stem) lest ye shall be defiled. 8. The whole city was moved because the dwellings had been shaken (VI stem). 9. He was anxious lest they should establish (cause to stand) a covenant with the city. 10. I died that ye might not die. 11. Put thy hand on his head and bless him before he die.

LESSON FOURTEEN. Gen. III. 6—14.

1. Notes.

234. ܕܫܰܦܺܝܪ *that* [*was*] *good.*

(1) ܕ introduces the object clause, § 135. 3.

(2) ܫܰܦܺܝܪ is the predicate, placed regularly and in the absolute state, § 99. 2.

(3) The clause is nominal, § 130. 1.

235. ܪܶܓܺܝܓ݂ܬܳܗܘ̱—*reg-geṯhau.*

(1) The ܗܘ is enclitic, and hence its ܗ is silent and its *u* coalesces with the preceding original *a* into *au* or *aw*, § 101, 23. 4.

(2) The noun has no absolute state. If it had we would expect to find it here. But see also, § 93. 2. (1).

236. ܠܡܶܚܙܳܐ *to see* or *for seeing.*

(1) The preposition takes *a* before the unvowelled consonant, § 34. 3.

(2) ܡܶܚܙܳܐ comes from *meḥwar* from *maḥwar; wa* going over regularly into *ô*, § 59. 1, § 29. 5. (3).

237. ܐܶܬ݂ܦܰܬܰܚ[ܝ] *were opened.*

(1) The final Yudh is found in some manuscripts and omitted in others. It is the sign of the fem. plural, § 43. 5.

(2) The Reflexive is used here as a Passive, § 41. 4.

(2) The Ḳushoy over the ܬ denotes doubling and hence the intensive stem, §§ 41. 2 and 10. 2. (2).

238. They made for themselves garments. For the construction of a verb with a direct and an indirect object, see § 125. 3. (2).

239. ܡܗܰܠܶܟ݂ [*as he was*] *walking.*

(1) The form is the intensive participle abs. sing., § 50. 2.

(2) The construction corresponds to the Ḥâl in Arabic, *e. g.* the accusative of condition, § 137. 7.

240. ܠܡܦܰܢܝܳܬܶܗ ܕܝܰܘܡܳܐ lit. *at the turnings of that which is day,* § 97. B.

241. ܐܶܬ݂ܛܰܫܝ[ܘ] *they hid themselves.*

(1) Some manuscripts omit ܘ.

(2) The masculine gender is preferred in the verb, when it has two subjects one masculine and the other feminine, § 121. 6. *Rem.* 2.

(3) The Reflexive sense is brought out clearly in this form.

242. ܚܙܝܬ for *hezyeth* from *hăz(ă)yith*, § 29. 4. (4), § 60. 1.

243. ܡܢܘ *who* [*is*] *he*. The form is a contraction of *man* and *hu*, § 39. *Rem.* 4.

244. ܚܘܝܟ *ḥaw-weyokh*. The verb is the Pa'el Perfect of the Lomadh Olaph verb. ܚܘܝ with the pronominal suffix of the 2nd masc. sing., § 61.

245. ܕܦܩܕܬܟ *which I commanded thee*.

(1) ܕ must be taken along with ܡܢܗ and translated "from which", § 104. 2. *Rem.*

(2) ܕ introduces a relative clause limiting ܐܝܠܢܐ, § 136.

(3) Paḳḳedhtokh is the intensive Perf. 1st. sing. with the pron. suff. 2nd masc. sing. § 51. 3.

246. ܐܢܬܬܐ *the woman* is resumed by ܗܝ. It stands in the nominative absolute, § 95. 3.

247. ܡܢܘ *what* [*is*] *he*. The form is contracted from *mônô* and *hu*, §§ 39. *Rem.* 4, 23. 4. The *hu* is here used as copula, the demonstrative limited by the relative making the predicate, § 101.

248. ܕܥܒܕܬܝ *which thou* (f.) *hast done*. The ending ܬܝ is derived from ܐܢܬܝ 2nd fem. sing. pers. pron., §§ 35. 1, 43.

249. ܐܛܥܝܢܝ—*'aṭ-e-yan*.

(1) The line under 'É is Mehagyono, § 12. 1.

(2) ܢܝ is the pron. suffix of the 1st sing. § 36. 1.

(3) The Olaph denotes the causative stem, § 42. 3.

(4) The Yudh shows that the verb is a Lomadh Olaph (Yudh) verb, § 60.

250. ܠܝܛܐ *cursed* is the pass. part. of the simple stem from the É Waw verb, *awî* going over into *î*, § 59. 3.

251. ܚܝܝܟ *thy lives*, §§ 36. 1.

2. Observations.

113. Nominal clauses are those which have a noun for predicate, *e. g.* the clauses beginning with ܕܫܦܝܪ vs. 6; ܕܥܪܛܠܝܢ vs. 7; ܐܝܟܐ

vs. 9; ܕܟܪܬܗܘܢ vs. 10, and vs. 11; ܡܢܗ vs. 13; ܒܣܝܡ vs. 14. Verbal sentences are those whose predicate and copula are a verbal form, *e. g.* the sentences beginning with ܘܚܙܬ vs. 6, ܢܩܦܘ vs. 7 etc.

114. As in Hebrew, the personal pronoun is frequently used as a copula, vss. 6, 11 and 13.

115. The personal pronouns may be used to emphasize the persons denoted by the forms of the verb, compare ܗܘ, vs. 12.

116. The Infinitive is really a verbal noun *i. e.* it is governed like a noun and governs like a verb, *e. g.* ܠܡܐܟܠ *for eating,* ܠܡܚܙܐ *for looking at,* vs. 6. See § 120.

117. The same forms are used to denote the Reflexive and Passives. For the former compare ܐܬܛܫܝܘ vs. 8, for the latter ܐܬܦܬܚܝ vs. 7.

118. The participle when used like the Arabic accusative of condition is indefinite but agrees with its antecedent in number and gender.

119. When the relative is to be governed by a preposition the relative ܕ is placed first as usual and the preposition follows with a pronominal suffix agreeing in gender and number and person with the antecedent of the relative.

120. Nouns are frequently placed in an abnormal position at the beginning of a sentence, their place in the sentence being assumed by a pronoun, *e. g.* ܐܢܬܬܐ vs. 12.

121. The relative time of the participles is to be gathered from the context. Compare ܡܗܠܟ vs. 8 with ܒܣܝܡ vs. 14.

3. Grammar.

(1) Guttural verbs, § 52.

(2) Pê Nun verbs, § 53.

(3) Review §§ 18, 26, 51.

4. Word Lists.

ܚܕܪ *to surround.*	ܣܢܩ *to want.*
ܙܪܥ *to sow.*	ܢܗܪ *to shine.*

ܢܗܰܪ *to be light.*
ܫܰܒܰܚ *to praise.*
ܐܶܫܟܰܚ *to be able.*
ܬܒܰܪ *to break.*
ܬܡܰܗ *to admire.*
ܢܛܰܪ *to keep.*
ܢܣܰܒ *to take.*
ܢܟܰܣ *to slay.*
ܢܦܰܩ *to go out.*
ܢܦܰܠ *to fall.*
ܢܦܰܨ *to scatter.*
ܐܶܡܳܐ *mother.*

5. Exercises.

1. Adam saw that the true was good to look at. 2. The mother of the man saw that the fruits of all the trees (were) good for eating. 3. Where (art) thou (f.), the desire of my eyes? 4. The husband of the woman gave some (ܡܶܢ) of the fruit to his father and to his mother and they saw (masc.) that the tree from which it was taken (which it was taken from it) was pleasant to the eyes of both of them and they ate and praised God, who had made them (ܡܶܢ) the dust. 5. Adam ate and knew that he was naked and he sewed the leaves of a figtree and made for him an apron. 6. My mother heard the Lord walking in the garden and she hid herself in the midst of a figtree which was in the garden. 7. At the turnings of the day I heard a voice saying, Adam, Adam, where art thou? And I saw God in my image and according to my likeness walking in the garden. 8. I said to the Lord: Who told thee that I and my wife are naked. Behold from the serpent hast thou heard this. 9. Surrounding, he shall take, keep thou (m.), go out (f.), it will fall, he caused to break, praise ye (Pa'el), it will shine, I shall not want, sow ye (m.). 10. I admire him because he was able to slay the Tanninin. 11. I cannot take my mother with me.

LESSON FIFTEEN. Gen. III. 15—24.

1. Notes.

252. ܒܥܶܠܕܒܳܒܽܘܬܳܐ *(the) enmity.*

(1) This is an abstract noun in ܘܬ, § 75. 4, derived from the compound word *b'eld'bhobho, enemy,* compound of ܒܥܶܠ *lord* and ܕܒܳܒܳܐ *fly,* § 96. 1. *a.*

(2) Most nouns with this ending are found only in the emphatic state, and are consequently often used when the idea is indefinite, § 93, 2. (1).

253. ܐܝܣܝܡ *will I put.*

(1) The Olaph is the sign of the first pers. sing. of the Imperfect, § 45. 5.

(2) The ܝ comes from *yi*, § 59. *Rem.* 2, § 29. 4. (4).

(3) This is the only 'Ê Yudh verb which differs in any respect from Ê Waw verbs, § 59. 6. *Rem.* 2.

254. ܢܕܘܫ from *nedh-wush*, *wu* becoming *û*, §§ 29. 7. (1), 59. 2.

255. ܬܡܗܝܘܗܝ—*tem-hêoo.*

(1) Notice the peculiar diphthong, pronounced like *ey* in *they* followed by *oo* as in *booby*, § 8. 1. (1).

(2) ܘܗܝ is the pron. suff. 3rd. masc. sing. This is the regular form after the vowel *e*, § 36, § 61.

(3) The first Yudh belongs to the root of the verb; the Taw is the preformative for the 2nd pers. Imperf., § 45.

256. ܡܣܟܢܘ is an Inf. abs. of the Aph'el stem, § 49. 2. It strengthens the idea of the verb, § 119.

257. ܬܐܠܕܝܢ *shalt thou bear.*

(1) The ܝܢ at the end is the sufformative of the 2nd fem. sing. of the Imperf., § 45.

(2) The root is ܝܠܕ, the Yudh beeing changed to Olaph after the preformatives of the Imperf. P^{e}'al, § 58. 2.

(3) The text has by mistake *e* for *î* under the preformative.

258. ܒܢܝܐ *sons*, is an irregular plural from ܒܪܐ son, §§ 86. 16, 87. 10.

259. ܬܬܦܢܝܢ *shalt thou turn thyself.* This is the Ethp^{e}'el Imperf. 2nd fem. sing. § 60 from ܦܢܐ.

260. ܢܫܬܟܠܠܬ—*neshtallat.* Note the transposition of the ܬ when before a sibilant, § 21. 1.

261. ܡܛܠ ܕ introduces the causal adverbial clause, which is nere the protasis; the apodosis beginning with ܗܟܝܠ, § 137. 5.

262. ܒܪܝܟܐ is in the fem. abs. sing. of the pass. participle. It is the predicate of the nominal clause of which ܐܢܬܝ is the subject, §§ 130. 1, 99, 2.

263. ܬܶܐܟܠܺܝܗ̇ *thou shalt eat [of] it.*

(1) The form of the verb when without the suffix is ܬܶܐܟܽܘܠ; with suffixes the ܽܘ is changed (volatilized) to a half-vowel, §§ 7. 1. (3), 7. 3. (1), 31. 1.

(2) After a vowel, the 3rd fem. pron. suffix is ܗ simply, which is often marked with a diacritical point over it, § 36 and § 6. 6. (2).

(3) The union vowel of the Imperf. 3rd sing. masc. and like forms is regularly ܺ before the pron. 3rd sing. masc. or fem. See § 51. D. 2.

264. ܬܰܘܦܶܩ *shall it bring out.*

(1) ܬ is the sign of the fem. 3rd sing., prefixed in the Imperfect, § 45. 2.

(2) The vowel ܰ with the preformative denotes the Causative stem, § 42. 5.

(3) The original Wau of Pê Wau verbs remains in the Aph'el, not passing over into Yudh as in the Pe'al, § 58. 1 nor contracting into ô as in Hebrew, § 58. 3.

265. ܕܡܶܢܳܗ̇ *which from it i. e. from which.* When the preposition governs a relative, the ܕ stands at the head of the sentence and the preposition comes after in the sentence followed by a pronominal suffix agreeing in gender and number with the antecedent of the relative, § 104. 2. *Rem.*

266. ܗ̣ܝ *hî, she* is put here for emphasis, § 101. The point under the ܗ shows that *hî* is to be read and not *hoy*, § 6. 6. (2) *b.*

267. ܕܚܰܝ *which [is] living.*

(1) This is really a complete relative sentence, of which ܕ is the subject and ܚܰܝ the predicate, the copula being unexpressed, § 136. 1. (1).

(2) ܚܰܝ is an adjective and agrees with its antecedent in gender and number, § 99. 2.

268. ܟܽܘܬܺܝܢܝܳܬܳܐ, sing. ܟܽܘܬܺܝܢܳܐ. A Yudh is inserted before the plural ending in a number of feminine nouns, § 86. 13.

269. ܐܶܢܽܘܢ *them.* There being no pron. suffix for the 3rd plural with verbs, the independent personal pronoun is used instead, § 36. 2.

270. ܐܺܝܕܶܗ *his hand.*

(1) Olaph is prosthetic, § 10.

(2) Ḥebhoṣo is a helping vowel, §§ 20, *Rem.* 2, 33. 1.

(3) For the irregularities of ܣܺܡ, see § 87. 2.

271. ܢܶܣܰܒ݂—*nessabh* for *nenṣabh*, the Nun being assimilated, §§ 18. 1, 53. 2.

272. ܢܺܚܶܐ *he shall live.* This is the Imperfect Pᵉ'al from ܚܝܳܐ. See § 64. 6.

273. ܘܰܐܦܶܩܶܗ—*wappᵉḳeh.*

(1) ܗ̄ is the pron. suffix 3rd masc. sing., § 36, 51. A.

(2) The Wau being unvowelled has drawn back the vowel of the Olaph the latter quiescing, §§ 34. 2, 25. 1. (2).

(3) The full form of ܐܦܶܩ was ܐܰܢܦܶܩ the usual Aph'el. The Nun has been assimilated, the ̄ has become a half vowel before the suffix, the ̱ has been thrown back to the Wau, § 53. 2.

274. ܡܰܕܢܚܳܬ݂ is a construct plur. before a clause beginning with a preposition, § 96. 4. *Rem.* 1.

275. ܕܡܶܬ݂ܗܰܦܟ݂ܳܐ *which was turning itself.*

(1) This is an adjective clause, the predicate being in the fem. abs. sing. agreeing with its antecedent, § 99. 2.

(2) The Rukhokh under the ܦ shows that this is the Ethpe'el, § 44, *Rem.* 1.

2. Observations.

122. The composition of two nouns to express one idea is occasionally found in Syriac.

123. The differentiations for gender, number and person in the verb are denoted by pre- and sufformatives.

124. Notice the importance of learning the contractions of Waw and Yudh with the vowels, *e. y.* in ܐܬܚܙܝ and ܢܡܘܬ.

125. There is a diphthong *eu* found in Syriac which is pronounced somewhat like Italian *eu* in *eufonia.*

126. The Infinitive is used absolutely to intensify the idea of a cognate verb which it precedes.

127. Instead of an adjective agreeing in definiteness with its antecedent, we frequently find a relative clause, *e. g.* ܕܚܰܝ.

128. There are a great many irregular plurals in Syriac which must be learned one by one. Compare ܒܢܝܐ, ܒܘܠܝܢܬܐ, §§ 86, 87.

3. Grammar Lesson.

(1) 'Ê Olaph verbs and Lomadh Olaph Guttural verbs, §§ 56, 57.

(2) Review §§ 55, 52, 24, 25, 26, 31, 32, 33.

4. Word List.

ܫܐܶܠ *he asked.*

ܒܐܶܫ *it was evil.*

ܟܐܶܢ *it grieved.*

ܣܐܶܒ *he was old.*

ܣܐܶܢ *he put on his shoes.*

ܛܐܶܒ *he was good.*

ܒܥܶܠܕܒܳܒܳܐ *enemy.*

ܝܗܰܒ *he gave.*

ܟܽܠ ܡܳܐ ܕ *all that.*

ܟܽܠ ܡܰܢ ܕ *whosoever.*

ܛܢܰܦ *he was filthy.*

ܒܰܝܺܐ *he consoled.*

ܛܰܡܺܐ *he was unclean.*

ܒܪ *a son.*

ܗܳܫܳܐ *now.*

ܐܶܢ *if.*

5. Exercises.

1. The enemy asked that my sword be given to him. 2. It grieved (fem.) me (ܠܝ) that I was too old to put sorrow for his bread. 3. All that was good to me was evil to him. 4. When a son was born to her she consoled herself. 5. Whosoever is filthy now, will be unclean all the days of his life. 6. If thou wilt crush my head, I shall strike thee in thy heel. 7. Thou didst command me that in the sweat of my face I should eat the herb of the field, until I shall return unto the dust from which I was taken. 8. Call the name of the woman Eve; because she shall be the mother of all which shall live. 9. God will make coats of skin for you and will clothe you. 10. Like one of you I know the good and the evil and I shall stretch out my hand and shall take from the tree of life and I shall live for ever. 11. The Lord sent them out from Eden that they might till the earth from whose dust they had been made by him. 12. The cherub turned itself and kept the way to Eden.

LESSON SIXTEEN. Gen. IV. 1—13.

1. Notes.

276. The point under the Nun in ܩܢܝܬ and under the Lomadh in ܝܠܕܬ and the ܣ in ܐܘܣܦܬ shows that these are the 3rd fem. sing.; the point above the Ḳoph in ܩܢܝܬ denotes the first peron singular, § 6. 6. (3).

277. ܠܡܐܠܕ, § 58. 2.

278. ܠܐܚܘܗܝ *his brother.*

(1) The Lomadh is the sign of the direct object, § 123.

(2) ܐܚܐ *brother*, and ܐܒܐ *father* insert ܘ before suffixes except the 1st sing., § 87. 1.

279. ܪܥܐ The point over the ʿÊ shows that this is a participle; a point under would denote a Perfect (Comp. ܐܡܪ vs. 10). It is either in construction with or governing, ܥܢܐ in the accusative, §§ 118. 2, 123. The dots over ܥܢܐ denote the collective, see § 90.

280. ܡܢ ܩܨܬ *after some.*

281. ܐܝܬܝ *he brought.* Aph'el Perf. 1st form from ܐܬܐ, § 64. 4.

282. ܐܬܪܥܝ, §§ 21. 1, 22. 4.

283. ܐܬܒܐܫ, §§ 56. 2, 25. 1. (2).

284. ܐܬܟܡܪܘ, §§ 52. 3, 43. A. The Rebbuy § 13 is put with this form to show that it is not a 3rd fem. sing. § 43. B. 5.

285. ܬܫܦܪ, §§ 41. 3, 42. 5, 45. B. 2, 52. 3.

286. ܐܢ *if* introduces the conditional protasis, § 138.

287. ܬܪܥܐ is of the one short vowel class of nouns, § 67. ܚܛܗܐ is of the ă—â class; ܪܒܥ of the ă—î class, § 69.

288. ܢܪܕܐ *let us go.*

(1) Remember that the preformative Nun denotes the 1st pers. plur. as well as the 3rd person, § 45. 10.

(2) The Imperfect is used for the 1st person of the Imperative, § 114. 1.

289. ܟܕ *when* introduces an adverbial clause of time, § 137. 2. The sentence is nominal, § 130. 1.

290. ܕܐܚܝ *of my brother.*

(1) When a noun in the genitive is separated from the noun on which it depends, the latter takes a pron. suffix agreeing in gender and number with the governed noun and the genitive is preceded by ܕ, § 97. B. *Rem*. 2.

(2) The vowel ̊ is heightened from ̄, § 7. 2. (4). See 277 above.

291. ܩܳܠܳܐ *the voice of the blood of him who is thy brother*, § 68. 5, 97. A. B.

292. ܕܬܶܬܶܠ *that it should give=to give.*

(1) *Tettel* is third fem. from *nettel* which is the singular Imperf. of ܝܰܗ݇ܒ, § 64. 7.

(2) The clause is an adverbial clause of result, § 137. 4 which is often expressed by the Infin., § 120. 1. (3) and see 276 above.

293. ܙܳܐܶܥ—*zo-yá*, *a fugitive.*

(1) The *a* instead of *e* is because of the guttural, § 26. 1. (1).

(2) The Olaph is inserted in the first form of the Participle of 'ÊWaw verbs, taking the place of the Yudh, § 59. 4. Compare the Ḥemsa in Arabic.

(3) This Olaph is pronounced like Yudh, § 2. (1).

294. . . . . ܪܰܒܳܐ lit. great is my folly from that which can be remitted *i. e.* my sin is too great to be remitted.

(1) The comparative degree is generally expressed by putting the adjective first in agreement with the noun to be compared and by placing the idea with which it is to be compared after the preposition ܡܶܢ, § 100.

(2) ܗ̱ܝ is here used as a copula, § 101. It is to be noted that the copula also agrees with the subject of the nominal sentence.

(3) ܣܟܠܘܬܝ *my folly* or *sin*. The ܝ is the 1st pers. pron. suffix, § 36; the *ûth* is the abstract fem. ending, § 75. 4.

(4) ܕܠܡܫܒܩ lit. *that which is to remit*. The ܕ is often used for *that which*, § 104. 2. (1). The Infin. preceded by Lomadh sometimes has the sense of "may" or "can", § 120. 1. (5).

2. Observations.

129. The distinction of forms as well as vowels by means of diacritical points is to be noted.

130. Every point and sign denotes *something*. The student is now far enough advanced not to proceed without knowing every verse thoroughly.

131. When a noun or verb is irregular, *i. e.* not according to the forms already learned, look in §§ 62—64, 86, 87.

132. Idioms should be carefully observed and if possible committed to memory, *e. g.* ܩܡ ܩܕܡ, vs. 3, ܐܘܫܛ ܠܡܠܟܐ, vs. 2.

133. Try to remember the euphonic changes such as permutation and transposition and assimilation.

134. Classify, if possible, every noun according to its original form. It gives accuracy, especially in reading unpointed texts.

135. Memorize all particles. It saves time to do so.

3. Grammar Lesson.

(1) 'Ê'Ê verbs, § 54.
(2) Read over the declension of nouns, §§ 78—85.
(3) Review, §§ 76, 77.

4. Word Lesson.

ܚܐ *to live.*
ܚܫ *to suffer.*
ܥܠ *to go in.*
ܪܓ *to covet.*
ܟܦ *to bend.*
ܪܬ *to tremble.*
ܚܦܦ *to cherish.*
ܛܠܝܘܬܐ *youth.*
ܣܒ *old.*
ܛܒ *good.*
ܐܬܪܘܪܒ *to be magnified.*
ܩܕܡ *before.*
ܩܕܡ ܕ *before that.*
ܫܛ *to reject.*
ܕܗܒܐ *gold.*
ܕܝܢ *judgment.*
ܛܠ V. *to overshadow.*
ܐܦܩ *to divulge.*
ܡܟ *to be humble.*
ܢܩܫ *to sound.*

ܡܢ ܫܠܝ *suddenly.*

ܪܕܘܦܝܐ *persecution.*

ܩܠܐ *voice.*

ܡܪ *to be bitter*, IV. *to be made bitter.*

ܡܫܝܚܐ *Messiah.*

5. Exercises.

1. I suffered persecution because I had divulged the judgment of God. 2. Go in and live in the land whose gold (which her gold) you have coveted. 3. They (fem.) trembled and bent their faces to the earth. 4. The good youth cherished his old father (his father the old) and his old mother. 5. Let God be magnified and let me humble myself before him. 6. Embittered (shall be) his spirit when the voice of God shall sound the judgment because he has rejected the Spirit of God which brooded over the face of the waters when God had created the heavens and the earth and overshadowed the mother of the son of God, before that she conceived and brought forth the Messiah. 7. Eve added to bear Abel the brother of Cain; and after some days Cain who tilled the soil (was working in the earth) brought as a gift to the Lord some of the fruits of his soil and the Lord looked not with favor on his gift, because he had not done well.

LESSON SEVENTEEN. Gen. IV. 14—28.

1. Notes.

295. ܐܦܩܬܢܝ—*appekton*, § 51. B. 3. Notice that the union vowel of the 2nd pers. masc. sing. with suffixes is *o*.

296. ܩܕܡܝܟ. Some prepositions take the plural form before suffixes, § 77. 4.

297. ܐܗܘܐ has the point above to denote the first person, § 6. 6. (3).

298. ܟܠ ܡܢ ܕ *every one who*, § 107. 7.

299. ܢܬܬܒܥ. The second vowel is added, § 33. 3.

300. ܚܕ ܫܒܥܐ *one for seven i. e.* seven fold.

301. ܬܪܬܝܢ *two.*

(1) The numbers one and two agree with their nouns in gender.

(2) For the position and date, sed § 110. 1.

302. ܠܒܢܘܗܝ. The preposition Lomadh sometimes denotes the genitive, § 98. 1.

303. ܕܐܚܝܕܝܢ *who hold.* This is one of the few passive participles which are used in an active sense, § 117. 4.

304. ܗܘ takes up and makes emphatic the ܠܐ which precedes, §§ 95. 3, 101.

305. ܚܬܗ *his sister.* An Olaph has been rejected from before the Heth, § 23. 1. (1).

306. ܫܡܥܝܢ is the 2nd fem. plur. of the Imperative in *a*, § 48. 2. ܨܘܬܝܢ is in the same place, § 59. 2.

307. ܐܚܪܢܐ *another.*

(1) The Olaph is occult, § 19. 1. (1) and hence is denoted by the linea occultans, § 11.

(2) Attributives follow their nouns and agree with them in gender, number and state, § 93. 3, 99. 1.

308. ܫܪܝ *began he (or they).*

(1) The dot above the Shin shows that the verb is Pa'el, § 6. 6.

(2) Either the subject is Seth, or the verb is impersonal, § 122.

2. Observations.

136. Pay attention to the union vowels of the different forms of the verb before the various suffixes.

137. Some prepositions take the plural, some the singular, form before suffixes, *e. g.* ܥܠ and ܩܕܡ take the plural form, ܥܡ and ܠܘܬ the singular.

138. There is a number of ways of expressing the indefinite pronoun in Syriac. The most common is to have the interrogative pronoun preceded by ܟܠ and followed by ܕ.

139. The rules for cardinal numbers are the same as in Hebrew.

140. Notice the fourth way of expressing the genitive relation, vs. 20.

141. Some participles which are passive in form are active in sense, *e. g.* ܐܚܝܕ, vs. 21.

142. In looking for the derivation of a word or for its equivalent in the cognate languages, always see first, if possible, whether a letter has been rejected or not, *e. g.* ܫܬܳܐ, vs. 22.

3. Grammar Lesson.

(1) Doubly Weak Verbs, § 62.

(2) Read over the classifications of nouns, §§ 66—75.

(3) Review § 61.

4. Word Lesson.

ܢܣܳܐ *to tempt.*

ܐܶܣܳܐ *to heal.*

ܢܳܚ *to rest.*

ܛܪܰܕ *to reject.*

ܚܰܘܺܝ *to show.*

ܟܳܐܐ *to rebuke.*

ܐܰܢ *to sigh.*

ܐܶܬܝܰܐܰܒ *to desire.*

ܐܶܬܐܰܘܺܝ *to agree*

ܐܶܬܪܰܓܰܓ *to desire.*

ܙܶܕܩܬܳܐ *alms.*

ܩܽܘܕܫܳܐ (m.) *holiness.*

ܚܕܽܘܬܳܐ *joy.*

ܩܽܘܪܝܬܳܐ *city.*

ܒܬܽܘܠܬܳܐ *virgin.*

ܒܥܽܘܬ *request.*

ܛܠܺܝܬܳܐ *girl.*

ܡܫܺܝܚܳܐ *Messiah.*

5. Exercises.

1. The Lord said to the girl I will heal thee and will put a sign on thee and cause thee to dwell in the land of Nod. 2. The girls tempted the Lord and he caused them to be rejected from the city of holiness. 3. He caused Cain to rest in the city which his son had built because he desired that he should not be killed. 4. The virgin will sigh when she sees thee because thou hast rebuked her and hast rejected her request. 5. A son has been born to the virgin and thou shalt call his name Messiah. 6. Be thou agreeing with him and do not reject his request. 7. Give alms to every one who asketh of thee and there shall be joy to thee.

LESSON EIGHTEEN. Psalm II.

1. Notes.

309. ܠܡܢܐ *why?* lit. *for what?*

(1) This is the adverbial accusative of cause.

(2) This is the common form of the neuter of the interrogative pronoun, § 39.

310. ܥܡܡܐ *the peoples,* § 86. 3. Singular ܥܡܐ.

311. ܪ̈ܢܝ is a fem. plur. of the Perf. from a Lomadh Olaph Verb, § 60. 1.

312. ܐܟܚܕܐ *together,* lit. *as one.* Note the insertion of the helping vowel *e*, § 33. 4.

313. ܕܝܬܒ *he who sitteth.*

(1) The relative ܕ sometimes stands for "he who", "that which" etc., § 104. 2. *Rem.*

(2) The participle denotes customary actions or a continuous state, § 116. 2.

(3) For the form, see § 99. 2.

314. ܐܩܝܡܬ from *aḳ-yimeth,* from *akwimeth,* § 59. 3.

(1) ܐ denotes the causative, § 41. 3.

(2) ܬ denotes the 1st person sing. of the Perfect, § 43. 5.

315. ܡܠܟܝ—*malk^e, my king,* §§ 36, 31. *Rem.* 1.

316. ܕܢܫܬܥܐ *that he may declare.*

(1) ܕ is a conjunction introducing the adverbial clause of purpose, § 137. 4.

(2) The verb is Ethp$^{e'}$el, § 41. 4, Imperfect, as shown by the preformative, § 45. It is determined as 3rd person sing. in distinction from the first plur. by the sense. Lomadh Olaph, § 60.

(3) The Shin and Tau have been transposed, § 21. 1.

317. ܕܒܪܝ—*dh^{e}bher^e.*

(1) ܕ introduces the quotation, § 135. 3. (3).

(2) For the pronunciation of the final Yudh, see § 31. *Rem.* 1.

318. ܫܐܠ § 56. 2, 25. 1. (2), 32. 3.

319. ܐܶܬܪܰܕܘ § 21. 1, § 30. 2. (1), § 24. 2, § 12. *Rem.*, § 11. *Rem.*, 48. 3.

320. ܕܶܚܶܠܬܳܐ *fear.* The second *e* is a helping vowel inserted to facilitate the pronunciation of the guttural, § 28. 2. (2).

321. ܘܶܐܚܽܘܕܽܘܗܝ.

(1) ܘܗܝ is the regular pron. suffix 3rd. sing. masc. after a verbal form of the plural ending in a consonant, § 51. A.

(2) The vowel *u* of the Imperat. is shifted before suffixes, § 51. E. and § 32. 1.

322. ܡܶܛܽܠ ܕ introduces the adverbial clause of cause, § 137. 5.

323. ܫܳܓܰܪ *burneth.* The participle denotes a state, or action viewed as continuing, § 116. 1.

324. ܕܬܳܟܠܺܝܢ those who trust, §§ 117. 4, 99. 2, 104. 2. *Rem.*

2. Observations.

143. Nouns and pronouns may, without any change of accidence, be treated as the Arabic adverbial accusative or the Latin oblique cases to denote cause, time etc.

144. Some irregular plurals are formed by inserting Wau before the regular ending, *e. g.* ܐܡܘܬܐ, vs. 1.

145. Compound words are occasionally met with, *e. g.* ܐܚܫܪܐ, vs. 2.

146. Remember the use of the relative ܕ in the sense of "he who" when followed by a participle vs. 3 and compare the Greek and Hebrew with the article.

147. Notice how often the Syriac translators have changed the Hebrew Imperfects, Perfects etc. into different tenses, *e. g.* the Hebrew Imperfects in vss. 1 and 2 have been changed into Perfects. Let the student hereafter note these changes and seek their cause.

148. Notice the light which a study of this psalm throws on the Syrian translators' views of the text, grammar and exegesis of the psalm, *e. g.* ܐܬܡܫܚܬ vs. 6, ܒܪܐ vs. 12 etc.

149. Notice the differences as well as the similarities between the Syriac and the Hebrew in root, form and construction, *i. e.* as to roots,

consider (1) sometimes the same root has a different meaning in the two languages, *e. g.* ܛܥܢ, ܠܘܬ, ܦܩܕ, ܡܠܟ.
(2) Sometimes the same idea has a different root, *e. g.* to forsake, to make, to form.

3. Grammar Lesson.

1. Anomalous and Defective Verbs, § 64.
2. Numerals, § 88.
3. Read, §§ 63, 65, 86, 87.
4. Review §§ 43, 45, 53, 54.

4. Word Lesson.

ܐܙܠ *to go.*	ܝܗܒ *to give.*
ܐܫܬܝ *to drink.*	ܣܠܩ *to ascend.*
ܐܫܟܚ *to find, to be able.*	ܘܠܐ *it behooves.*
ܐܬܐ *to come.*	ܝܐܐ *it is well.*
ܗܘܐ *to be.*	ܬܪܝܨ *it is right.*
ܚܝܐ *to live.*	ܟܪܐ *to grieve.*
ܠܥܠܡ *ever.*	ܡܐܢ *to be weary with.*

5. Exercises.

1. Go thou and see why the three rulers have taken counsel together against the Lord and against his Messiah. 2. The Lord will give Zion the mountain of his holiness to his son the king. 3. Who shall ascend to the mountain of the Lord? Who shall be able to stand in the place of his holiness? 4. Let the king live for ever; let the peoples come and serve him because it is right for them to serve him with fear. 5. It behooves us to kiss the son lest he be angry and we perish from his way because that his wrath has been kindled against us. 6. The Lord was weary with the two peoples because they imagined a vain (thing) and said: Let us break the bands of the Lord and cast from us his yoke. 7. It grieved the four kings that they should not be for ever. 8. It is well to drink water from the fourth vessel of the eighth potter.

PART II.

Notes.

Jonah I.

325. ܠܡܐܡܪ *saying.* For the idiom compare the Hebrew and see § 120. 1. (3) and note 130. (2).

326. ܙܠ "*Go*". Imperative from ܐܙܠ, § 64. 1.

327. ܡܕܝܢܬܐ—*m'dhîto, city,* § 18. 2.

328. ܥܠܝܗ *against her.* ܥܠ takes a plural form before suffixes, § 77. 4, as also ܩܕܡ.

329. ܣܠܩܬ, § 64. 8.

330. ܐܫܟܚ, § 64. 3.

331. ܐܙܠܐ Part. act. fem. from ܐܙܠ, § 54. 3. ܡܐܙܠ is the Infin. of the same.

332. ܪܒܬܐ. Notice the position of the adjective after its noun and its agreement with it in gender, number and state, § 91. 1.

333. ܐܢܫ *each.*

(1) The Olaph is occult, § 19. 1.

(2) For the use of ܐܢܫ for the indefinite pronoun, see § 107. 2.

(3) ܐܢܫ when denoting each or every one takes a plural verb, § 121. 2. See further, § 90. 4. *Rem.* 2.

334. ܢܬܚܢܢ V *stem.* 'Ê'Ê verb, § 54.

335. ܠܗ line 8 is an ethical dative, §§ 124. 5, 101. B. 1. (1) *Rem.* 3.

336. ܢܦܨܝܢ l. 10. III. stem Imperf. 3rd sing. with pron. suff. 1st plural, § 61. 2. ܝ is contracted from *ay*, § 29. 3. (1).

337. ܬܘ l. 11, *come.* Imperat. from ܐܬܐ, § 64. 4.

338. ܢܕܥ—*nedda', let us know,* 1st pers. plur. Imperfect I stem from ܝܕܥ, § 58. 2. *Rem.* 1.

339. ܚܘܢ l. 12, *show thou us.* III stem Imperat. sing. masc. with pron. suff. 1st plur., § 61. 3.

340. ܡܢܘ l. 13, *what is?* § 103. 1. (1), § 39. *Rem.* 4, § 23. 4. (1).

341. ܐܝܢܐ *what?* § 39. *Rem.* 3, § 103. 2. (2) *Rem.* This is an interrogative adjective separated from its noun by the personal pronoun.

342. ܕܚܠܘ p. 12, l. 1, §§ 34. 2, 33. 1.

343. ܓܒܪܐ *the men.* Rebbuy denotes the collective, § 90. 1. See also 333 above.

344. ܢܬܚܡ, § 46. 1.

345. ܘ *that,* § 137. 4. (1).

346. ܫܩܘܠܘܢܝ Imperat. 2nd masc. plur. with pron. suff. 1st sing., §§ 51, 36. 1, 32. 1.

347. ܐܪܡܘܢܝ l. 3. § 61. 1. (3).

348. ܡܛܠܬܝ l. 4, *on account of me.* The preposition ܡܛܠ takes the fem. plur. form before suffixes, § 89. B. (6). The ̈ is written with the Yudh, but belongs to the ܗܘ following, the Hê having become occult because the pronoun is enclitic, § 19. 2. (4). Since a vowel cannot begin a syllable, the last consonant of the preceding word draws to it the vowel of the Hê, § 16. 2. If the preceding word end in a vowel, it forms a diphthong with the *u.* Compare ܐܢܐ ܗܘ 22. 15. The same is true of ܗܝ. Compare ܛܘܪܐ ܗܝ 22. 12.

349. ܗܢܘܢ, §§ 37. 2, 102. 1, 90. 1.

350. ܐܫܬܩܘ l. 6, §§ 20. *Rem.* 1, 64. 3.

351. ܐܪܓܠܗܘܐ, §§ 19. 2. (1) *a*, 64. 1, 116. 1. (3) *a.*

352. ܠܐ ܬܫܦܘܟ, § 115. 3. The Syriac does not distinguish between "thou shalt not" and "do not".

353. ܫܩܠܘܗܝ ܠܝܘܢܢ. *They took Jonah.* For the use of the pron. suffix to emphasize the object, see § 123. 2. (5). (6). (7). (8).

354. ܕܚܠܬܐ. A cognate accusative. See § 126. 4. (1).

Jonah II.

355. ܙܡܢ. III stem, §§ 56. 4. *Rem.* 59. 5.

356. ܨܠܝܘ ܠ, §§ 51. 1. 6, 123. 2. (6).

357. ܕ ܡܫܒܚܬܗ, §§ 13, 33. 2, 34. 3, 77, 97. B

358. ܬܟܠܬܐ l. 13, § 110, 1. (1).

359. ܟܢܫܝܗܘܢ, § 87. 19.

360. ܚܦܘܢܝ, §§ 77, 82. *Rem.* 7.

361. ܚܕܪܘܢܝ, § 61. 1, 36. 1.

362. ܥܒܪܘ. Some verbs in Syriac, as in other languages, take a preposition before their object, § 123. 5.

363. ܟܠܗܘܢ ܡܫܘܒܪܝܟ *all thy waves* (all of them, thy waves), § 94. 6. (1).

364. ܥܝܢܝܟ, § 12. 1.

365. ܕܐܬܛܪܕܬ. The relative introduces the quotation, § 135. 3. (3). The stem is here reflexive, § 41. 4.

366. ܡܬܢܩܦ. V stem Part. from ܢܩܦ, § 58. 3.

367. ܡܘܟܪܝܗ ܥܠܐܦܝ *the earth laid hold with its bands on my face, i. e. on me*, § 105. 1. (3).

368. ܐܣܩ, § 64. 8.

369. ܚܝܝ—*ḥay-yay, my life.*

370. ܐܬܕܟܪܬ. For the reflexive verb with an object, see § 126. 2. (1).

371. ܗܝܟܠܟ ܩܕܝܫܐ *thy holy temple.* Notice that the pronoun follows the noun and not the adjective, § 99. 1. *Rem.* 3.

372. ܡܢ ܕ *whoever*, § 107. 7, 8, and § 108. 2.

373. ܡܕܡ ܕ *whatsoever*, § 109. 1. (3).

Jonah III.

374. ܕܬܪܬܝܢ ܙܒܢܝܢ, § 110. A. 1. (1), B.

375. ܟܕ ܐܡܪ *saying*, § 120. 1. (3).

376. ܟܪܘܙܘܬܐ. For the form, *see* §§ 71. 1, 75.

377. ܪܒܬܐ ܠ *great to, i. e.* the greatest city, § 100. 2. (5).

378. ܡܬܗܦܟܐ *shall be overturned.* The participle is defined as future by the ܥܕܢܐ ܠ, §§ 111. 3, 116. 1. (2) *b.*

379. ܐܢܫܝܗ, lit. *her men*, §§ 19. 1. (1), 77.

380. ܠܒܫܘ *they clothed themselves with*, § 126. 2. (1). *Rem.*

381. ܪܘܪܒܢܝܗܘܢ *their magnates.* The Singular is ܪܒܐ, see § 87. 27. For the helping Rebhoṣo, see § 33. 3, 9. *Rem.*

382. ܟܘܪܣܝܗ *his throne*, § 86. 2. (2).

383. ܒܢܝ ܐܢܫܐ *the sons of men*, §§ 87. 10, 23. 4. (1).

384. ܒܚܝܠܐ, § 90.

385. ܡܕܡ *anything*, § 109. 1. (1).

386. ܢܩܪܘܢܗ ܠܐܠܗܐ *let them call God*, § 123. 2. (7).

387. ܐܢܫ. Each, § 107. 2.

388. ܚܛܘܦܝܐ, § 81. *Rem.*

389. ܕܐܝܬ *which is*, §§ 65, 128. 3. (2).

390. ܒܐܝܕܝܗܘܢ, §§ 87. 2, 20. *Rem.* 2, 34. 2.

391. ܡܢ, §§ 39. 1. *Rem.* 1, 103. 1.

392. ܡܢܘ, §§ 116. 1. *a*, 52. 3, 26. 1. (1).

393. ܐܢ introduces the indirect question, § 132. 6. (1).

394. ܕܠܐ introduces the negative adverbial clause of result, § 137. 4.

395. ܕܗܦܟܘ *that they turned.* This is an appositional substantive clause, § 135. 5.

Jonah IV.

396. ܘܒܐܫ ܠܝܘܢܢ. *It was painful for Jonah.* See § 122. 2.

397. ܛܒ *very*, is a masculine noun in the absolute state used as an adverb, § 89. A.

398. ܠܐ ܗܘܬ *was not?* The answer "yes" is expected, though ܠܐ itself does not denote this § 132. 2. A question is often denoted in Syriac without any interrogative particle or pronoun, § 132. 1. sq.

399. ܟܕ ܐܢܐ *when I (was)*, § 130. 1. (1).

400. ܩܕܡܬ ܗܘܝܬ *I anticipated*, § 127. 1, 3. (1) *a*.

401. ܥܪܩܬ *I fled.* With the preceding verb this verb may be translated "I fled before-hand", § 133. 3 and *Rem.*

402. ܠܝ is the Ethical dative or object, § 124. 5.

403. ܝܕܥ ܗܘܝܬ, §§ 127. 1, 116. 1. (3).

404. ܢܓܝܪܐ ܪܘܚܟ *long is thy spirit, i. e.* patient.

405. ܣܓܝܐܐ—*saggiyo'* from *saggi'o'*, §§ 24. 1, 32. 3.

406. ܫܩܘܠ, § 53. 1, 23. 1. (3).

407. ܢܦܫܝ, § 122. 4. (2) *Rem.*

408. ܠܡܡܬ *to die*, is the subject of the nominal sentence, § 120. 1. (1), § 130. 1. (1).

409. ܡܢ ܕܠܡܚܐ *than to live.*

(1) For the form ܡܫܝܚܐ, see § 64. 6.

(2) For the construction, see § 120. 1. (6), 100. 1. *Rem.* 2.

410. ܢܦܩ. See 398.

411. ܠܗ. See 402.

412. ܠܬܚܬܘܗܝ *under it*, § 89. B. (3).

413. ܕܢܚܙܐ *that he might see*, §§ 114. 4. (2), 137. 4.

414. ܡܟܝܠ, §§ 132. 6. *Rem.*, 135. 3. (2), 113.

415. ܢܓܕܫ *should happen*, § 116. 1. (3) *b*, 5.

416. ܩܪܐܐ *cucumber*, § 24. 1, 25, 28. 2. (3).

417. ܐܪܟܢ, §§ 26. 1. (1), 59. 6.

418. ܗܘ ܥܡܘܪܐ, 101. 3. (2) *a*.

419. ܠܢܦܫܗ for himself, § 105. 1. (3).

420. ܡܛܝܐ *it has come into thy hands, oh Lord, to take away my soul from me.*

421. ܡܛܠ ܕ *because that*, §§ 6. 5, 137. 5. (2).

422. ܠܐ ܗܘܝܬ § 127. 1. (2).

423. ܢܦܩ ܐܢܐ, § 99. 2. *Rem.* 1.

424. ܡܢ ܐܘܟܡܘܬܐ, § 100. 1, 87. 1, 86. 14.

425. ܐܝܠ. Emphatic, § 101. 1. (2).

426. ܕܠܐ—ܒܗ *on which—not*, § 104. 2. *Rem.*

427. ܐܢܐ. See 425.

428. ܠܩܕܡ ܡܢ, § 100. 1.

429. ܐܪܒܥܣܪܐ *fourteen*, § 88. 1, 100. A. 1. (4).

430. [illegible], § 85.

Malachi I.

431. ܐܫܬܥܒܕܬܘܢ, §§ 43. 5. *Rem.* 2, 51. A, B.

432. ܐܡܪܬܘܢ ܐܝܟܢܐ—*omrittun*, § 35. 2.

433. ܕܠܡܐ followed by ܠܐ expects the answer "yes", § 132. 5.

434. ܠܡܠܟܘܬܟܘܢ. The direct object may be preceded by Lomadh, § 123. 2.

435. ܘܐܢ *and if*, § 138. 2. (3).

436. ܢܬܪܥ from ܪܥ.

437. ܡܝܩܪ *is wont to honor*, § 116. 2.

438. ܐܠ ܐܦ̈ܐ, § 138. 4. (4).

439. ܕܡܩܪܒܝܢ, § 135. 5.

440. ܕܫܳܛܝܢ ܐܢܬܘܢ *ye who despise*, § 136. 11.

441. ܡܛܠ ܕ, § 137. 5.

442. ܒܐܡܪܝܢ, § 137. 5.

443. ܕܚܡܝܠ, § 135. 3. (3).

444. ܗܘ, § 101. 2.

445. ܩܪܶܒܝܗܝ, § 29. 2. (3).

446. ܕܢܚܡܐ, § 132. 4.

447. ܘܢܪܚܡ *that he may have mercy*, §§ 137. 4. (1), 114. 4. (2) *Rem.* 1.

448. ܕܡܢ ܐܝܕܝ̈ܟܘܢ *because this was in your hands*, §§ 104. 7.

449. ܡܩܒܠ ܐܢܐ, § 130. 1. *Rem.*

450. ܕܡܣܝܢ *that which is of no account.*

451. ܠܐ ܨܒܐ ܐܢܐ ܒܟܘܢ *I wish nothing among you.*

452. ܐܢܬܘܢ, vs. 12, is used as a copula, § 101. 2.

453. ܕܡܝܬܝܢ *because ye are bringing*, §§ 137. 5. (1).

454. ܠܝܛ ܗܘ ܡܢ ܕܐܝܬ ܠܗ *accursed be whosoever has*, § 107. 7. (4), 103. 1. *Rem.* 4.

Malachi II.

455. ܐܢ, vs. 2. § 138. 2. (2).

456. ܡܣܡ is infinitive from ܣܳܡ, § 64. 7.

457. ܡܛܠ ܕ, § 137. 5. (2).

458. ܐܕܪܐ read ܐܕܪܐ *I will scatter.*

459. ܐܢܘܢ, vs. 5. § 123. 2. (6.)

460. ܫܐܠܝܢ *they are asking*, § 121. 7.

461. ܠܣܓܝܐܐ *many.* The direct object is often preceded by Lomadh § 123. 1. (3).

462. ܡܫܡܫܝܢ, § 116. 3. (2) *c.*

463. ܕ *because*, § 137. 5. (1).

464. ܕܬܚܡ, §§ 46. 1, 136. 1. (2) 3.

465. ܕܡܩܪܒ *he who offers*, § 104. 2. (2) *Rem.*

466. ܕܪܘܚܗ, vs. 15, § 106.

467. ܕܒܝܫ *"that which is evil"*, is an objective clause, § 135. 3.

468. ܕܕܐܢ *who is judging*, § 116. 3. (1) *a*.

Malachi III.

469. ܗܐ *behold* is followed here by the Participle in the future.

470. ܐܢܐ ܡܫܕܪ ܐܢܐ, § 95. 1, 101. A. 2.

471. ܘܢܦܢܐ *that he may prepare*, § 137. 4.

472. ܕ . . . ܠܗ *whom*, § 104. 2. *Rem.*

473. ܡܬܒܥܝܢ, § 129. 2. *a*.

474. ܐܡܬܝ ܕ *when*, § 137. 2. (1).

475. ܕܡܢ *which were from*.

476. ܥܠ ܐܢܬܐ *against (him) who is turned to me*, § 103. 2. (3) *Rem.*

477. ܡܛܠ ܕܐܢܐ ܐܢܐ *because that I am*, § 101. 2. (1), 130. 1, 137. 5.

478. ܐܬܦܢܘ, vs. 7, § 60. 4.

479. ܢܣܐܘܢܝ, vs. 10, *prove me*, § 61.

480. ܡܠܝ, vs. 13, is masc. plur. the Wau being omitted, see §§ 23. 1, 43. 5.

481. ܘܡܬܒܢܝܢ *and are built up the doers of sin and (they) tempt God and are delivered.*

482. ܐܢܫ *a man with his neighbor, i. e. one with another.*

483. ܕܝܠܝ *mine*, § 106. *Rem.* 4.

484. ܕܚܐܣ, § 116. 2. (1).

485. ܕܦܠܚ *those who serve*, § 104. 2. (2). *Rem.*

Malachi IV.

486. ܕܝܩܕ *when shall burn*, § 137. 2. (1).

487. ܠܟܘܢ *to you, i. e. to the fearers of my name*, § 94. 1.

488. ܐܬܕܟܪܘ *remember*, §§ 11. 5. *Rem.*, 48. 3, 126. 2. (1).

489. ܠܟܘܢ ܠܐܠܝܐ *to you Elias*, § 124. 3.

Matthew XXVI.

490. ܐܡܪܝܢ ܗܘܘ, § 116. 3.

491. ܣܓܝ before its noun, § 99. 1. *Rem.* 1, § 96. 2. *b.*

492. ܕܡ̈ܝܐ an irregular plural used in a singular sense, § 86. 16.

493. ܐܬܚܡܬܘ, § 122. 5.

494. ܡܠܐܝܢ from ܠܐܝ *to trouble, to weary.* For the form see § 32, 3, 29. 1. (3).

495. ܠܗ̇ ܠܐܢܬܬܐ, § 123. 2. (7). *Rem.*

496. ܕܐܪܡܝܬ introduces the appositional substantive clause, § 135. 5.

497. ܐܝܟ ܕ *as that which is for my burial.*

498. ܠܕܘܟܪܢܗ̇ *for a memorial of her,* § 96. I. 4. *b.*

499. ܕܢܫܠܡ, § 129. 2. (3).

500. ܠܗ, § 124. 5.

501. ܚܕ ܚܕ *one by one.*

502. ܗܘ, emphatic, § 101. A. 1. (2).

503. ܠܘ, Impossible condition, § 138. 5.

504. ܐܢܐ ܗܘ, § 101. A. 2. (2).

505. ܗܢܘ ܕܡܝ ܕ *this is my blood that of the knew testament,* § 96. II. *Rem.* 1.

506. ܚܕܬܐ, § 87. 15.

507. ܠܓܠܝܠܐ, § 90. 4. *Rem.* 1.

508. ܐܦܢ *although,* § 137. 6.

509. ܐܢ, vs. 35, § 138. 2. (2).

510. ܬܢܢ ܠܟ, § 127. 9.

511. ܟܕ ܕܡܟܝܢ *sleeping,* lit. *while they were sleeping.*

512. ܐܠܐ . . ܕܠܟ, vs. 42, § 133. 1. *Rem.* 1

513. ܠܗ ܠܡܠܬܐ *the same word,* § 102. 3. (2) *a.*

514. ܗܘ ܕ, vs. 46, § 102. 7.

515. ܟܕ, vs. 47. *while,* § 137. 2. (2).

516. ܝܗܒ ܗܘܐ *had given,* § 127. 1. (2).

517. ܠܗܘ, vs. 48, *him whom I shall kiss, the same is he, him seize.*

518. ܠܗܕܐ, vs. 50, *is it on account of this thas thou hast come my friend?* §§ 132. 2, 135. 1.

519. ܗܘ̈, vs. 50, § 93. II. 1. (3).

520. ܕܝܬܒ, vs. 64, *him who sitteth,* § 104. 2. (2) *Rem.*

521. ܒܗ̇ ܒܫܥܬܐ *in the same hour,* § 102. 3. (2) *a.*

Matthew XXVII.

522. ܠܢ ܡܢܐ ܠܢ *what is that to us?* § 103. 1. Rem. 1.

523. ܐܙܠ, § 101. 2. (1).

524. ܡܕܡ, § 109. 1. (3).

525. ܕܢܦܫܗ, limits the preceding pronominal suffix, § 136. 4.

526. ܡܢ, vs. 9, *some,* § 107. 4.

527. ܟܕ, vs. 12, *while they were eating the pieces of him, i. e. calumniating him.*

528. ܟܕ, § 110. 1. (2).

529. ܕܝܠܗ, § 106. 1.

530. ܠܐ, vs. 19, *let there be nothing belonging to thee and to that just man, i. e. have thou nothing to do with him.*

531. ܟܠܗܘܢ, vs. 22. § 95. 2. (3).

532. ܗܘܐ, vs. 24, § 116. 1.

533. ܡܫܩܝܢ, Paʿel pass. part., § 60. 5.

534. ܐܡܪܘ, the plural verb, because the singular noun is collective, § 90.

535. ܠܟܠܗܘܢ, § 125. 1. (2).

536. ܡܚܝܢ ܗܘܘ, vs. 30, *kept smiting,* § 116. 2. (3).

537. ܟܕ, vs. 31, *as they were going out they found,* § 116. 1. (3) *b.*

538. ܟܕ, vs. 35, *"and when they had crucified him they parted his garments".* Notice the distinction between the Perfect here and the Participle in the preceding note, § 112. 1. (3).

539. ܐܚܝ *he made alive,* § 64. 6.

540. ܟܠ, § 110. 1. (1). Compare for ܟܠܗ, § 110. 1. (2).

541. ܐܦܝܗܘܢ, vs. 51. The first is in the masc. plur. agreeing with "faces"; the second is in the fem. plur. agreeing with "rocks". For the fem., see § 43. 5.

542. ܘܕܥܡܗ *and those who were with him,* § 104, 2. (2) *Rem.*

543. ܕܗܘ̈ܝܢ, §§ 60. 1, 127. 2, (2).
544. ܐܬܚܙܝܘ, § 127. 1. (2), 128. 2. *Rem.*
545. ܕܡܓܕܠ, vs. 56. § 96. II. *Rem.* 1.
546. ܗܘ, vs. 57, § 101. A. 1. (1).

Matthew XXVIII.

547. ܚܕ ܒܫܒܐ *the first day of the week*, § 88. *Rem.* 5.
548. ܐܬܬܙܝܥܘ ܗܘܘ, Compare 544.
549. ܕܕܚܠܬܗ, *fear before him.* Objective genitive, § 96. I. 4.
550. Vs. 5. ܢܫ̈ܐ *women*, § 87. 8.
551. ܕܐܙܕܩܦ *who was crucified*, § 136. 1. (2), 2.
552. Vs. 6. ܬܐܝܢ, ܚܙܝܝܢ, vs. 7. ܐܙܠܝܢ, §§ 60, 64. 1, 4.
553. ܩܕܡ, vs. 7, is Paʿel Perfect.
554. ܐܙܠܢ, vs. 8, fem. plur. part.; ܢܐܡܪܢ, Imperf. 3rd. fem. plur.
555. ܐܢ, § 138. 1. (3), 112. 3. (2) *b.*
556. ܕܠܐ, vs. 14, *those who are without care*, §§ 93. 2. (2), 104. 2. (2) *Rem.*
557. ܟܠ ܡܐ ܕ, vs. 20, *whatsoever*, § 109. 8.
558. ܦܩܕܬܟܘܢ, § 108. 1. (4).

WORD LISTS—SYRIAC.

LIST I.

Verbs occurring ten times or more in Schaaf's Concordance of the New Testament.

1.	ܐܒܕ	18.	ܒܐܫ	35.	ܒܬܪ	52.	ܕܠܐ
2.	ܐܒܠ	19.	ܒܕܪ	36.	ܓܒܐ	53.	ܕܥܪ
3.	ܐܓܪ	20.	ܒܗܬ	37.	ܓܙܪ	54.	ܕܩܪ
4.	ܐܙܠ	21.	ܒܙܚ	38.	ܓܕܦ	55.	ܕܠܫ
5.	ܐܚܕ	22.	ܒܛܠ	39.	ܓܠܐ	56.	ܕܟܪ
6.	ܐܚܪ	23.	ܒܠܥ	40.	ܓܥܪ	57.	ܕܪܝ
7.	ܐܝܬ	24.	ܒܢܐ	41.	ܓܢܒ	58.	ܕܪܫ
8.	ܐܟܠ	25.	ܒܥܐ	42.	ܓܡܠ	59.	ܗܘܐ
9.	ܐܠܦ	26.	ܒܥܠ	43.	ܕܒܚ	60.	ܗܠܟ
10.	ܐܠܨ	27.	ܒܨܐ	44.	ܕܒܪ	61.	ܗܦܟ
11.	ܐܡܢ	28.	ܒܩܐ	45.	ܕܚܠ	62.	ܗܪܓ
12.	ܐܡܪ	29.	ܒܣܪ	46.	ܕܝܢ	63.	ܗܘܠ
13.	ܐܣܐ	30.	ܒܟܐ	47.	ܕܟܐ	64.	ܙܟܐ
14.	ܐܣܪ	31.	ܒܪܐ	48.	ܕܟܠ	65.	ܙܕܩ
15.	ܐܪܟ	32.	ܒܪܪ	49.	ܕܡܐ	66.	ܙܗܪ
16.	ܐܫܕ	33.	ܒܣܐ	50.	ܕܡܪ	67.	ܙܠ
17.	ܐܬܐ	34.	ܒܙܐ	51.	ܕܢܚ	68.	ܙܩܐ

69.	ܐܚܪ .	98.	ܚܫܚ	127.	ܟܐܐ	156.	ܡܠܠ
70.	ܐܢܐ	99.	ܚܫܒ	128.	ܟܗܢ	157.	ܡܢܐ
71.	ܐܦܦ	100.	ܚܬܡ	129.	ܟܠܐ	158.	ܡܪܕ
72.	ܐܙܠ	101.	ܚܬܪ	130.	ܟܢܫ	159.	ܡܫܚ
73.	ܚܒܒ	102.	ܛܟܡ	131.	ܟܣܐ	160.	ܢܒܐ
74.	ܚܒܠ	103.	ܛܡ	132.	ܟܦܣ	161.	ܢܒܚ
75.	ܚܒܫ	104.	ܛܠܐ	133.	ܟܦܢ	162.	ܢܒܥ
76.	ܚܓܒ	105.	ܛܠܡ	134.	ܟܦܪ	163.	ܢܕܪ
77.	ܚܓܪ	106.	ܛܠܢ	135.	ܟܪܐ	164.	ܢܚ
78.	ܚܘܒ	107.	ܛܥܢ	136.	ܟܪܙ	165.	ܢܚܬ
79.	ܚܘܐ	108.	ܛܥܐ	137.	ܟܪܟ	166.	ܢܛܪ
80.	ܚܝܠ	109.	ܛܦܦ	138.	ܟܫܠ	167.	ܢܟܣ
81.	ܚܣ	110.	ܛܦܠ	139.	ܟܬܒ	168.	ܢܟܦ
82.	ܚܘܪ	111.	ܛܦܫ	140.	ܟܬܪ	169.	ܢܣܒ
83.	ܚܙܐ	112.	ܛܪܐ	141.	ܟܬܫ	170.	ܢܣܒ
84.	ܚܙܦ	113.	ܝܒܠ	142.	ܠܐܒ	171.	ܢܦܫ
85.	ܚܛܐ	114.	ܝܕܥ	143.	ܠܒܒ	172.	ܢܦܠ
86.	ܚܛܦ	115.	ܝܗܒ	144.	ܠܒܫ	173.	ܢܦܩ
87.	ܚܝܐ	116.	ܝܠܕ	145.	ܠܘܐ	174.	ܢܨܒ
88.	ܚܟܡ	117.	ܝܠܐ	146.	ܠܚܡ	175.	ܢܩܦ
89.	ܚܠܒ	118.	ܝܠܦ	147.	ܠܚܫ	176.	ܢܫܡ
90.	ܚܡ	119.	ܝܢܩ	148.	ܠܟܕ	177.	ܢܫܩ
91.	ܚܠܦ	120.	ܝܡܝ	149.	ܠܥܣ	178.	ܢܬܠ
92.	ܚܦܪ	121.	ܝܡܐ	150.	ܠܦ	179.	ܣܒܒ
93.	ܚܦܛ	122.	ܝܩܕ	151.	ܡܚܐ	180.	ܣܒܠ
94.	ܚܙܝ	123.	ܝܩܪ	152.	ܡܛܐ	181.	ܣܒܪ
95.	ܚܪ	124.	ܝܩܗ	153.	ܡܟܟ	182.	ܣܓܕ
96.	ܚܪܒ	125.	ܝܬܒ	154.	ܡܠܐ	183.	ܣܓܝ
97.	ܚܪܫ	126.	ܝܬܪ	155.	ܡܠܟ	184.	ܣܗܕ

185. ܣܡ
186. ܣܝܦ
187. ܣܒܐ
188. ܣܒܠ
189. ܣܠܐ
190. ܣܓܕ
191. ܣܥܪ
192. ܣܢܐ
193. ܣܠܩ
194. ܣܒܪ
195. ܣܦܩ
196. ܣܪܩ
197. ܣܗܕ
198. ܥܒܕ
199. ܥܒܪ
200. ܥܕܪ
201. ܥܗܕ
202. ܥܡ
203. ܥܙ
204. ܥܛܦ
205. ܥܠ
206. ܥܡܕ
207. ܥܡܠ
208. ܥܡܪ
209. ܥܢܐ
210. ܥܪܩ
211. ܥܗܕ
212. ܦܫ
213. ܦܓܥ
214. ܦܠܚ
215. ܦܢܐ
216. ܦܫ
217. ܦܫܩ
218. ܦܬܠ
219. ܦܓܪ
220. ܦܩܕ
221. ܦܘܚ
222. ܦܪܚ
223. ܦܪܣ
224. ܦܪܩ
225. ܦܪܥ
226. ܦܬܚ
227. ܦܣܩ
228. ܦܣܚ
229. ܨܒܐ
230. ܨܒܠ
231. ܨܘܡܐ
232. ܨܕ
233. ܨܡ
234. ܨܝܕ
235. ܨܠܐ
236. ܨܥܪ
237. ܨܘܪܐ
238. ܩܒܠ
239. ܩܒܪ
240. ܩܪܒ
241. ܩܕܫ
242. ܩܡ
243. ܩܛܠ
244. ܩܢܐ
245. ܩܥܐ
246. ܩܝܐ
247. ܩܪܐ
248. ܩܪܒ
249. ܩܡܐ
250. ܪܥܐ
251. ܪܗܛ
252. ܪܗܛ
253. ܪܚܡ
254. ܪܘܪܐ
255. ܪܒܥ
256. ܪܗܒ
257. ܪܗܛ
258. ܪܘܫ
259. ܪܡܐ
260. ܪܡ
261. ܪܫܡ
262. ܪܚܛ
263. ܪܟܒ
264. ܪܥܝܐ
265. ܪܢܐ
266. ܪܟܐ
267. ܪܩܕ
268. ܫܐܠ
269. ܫܒܐ
270. ܫܒܗܪ
271. ܫܒܚ
272. ܫܒܩ
273. ܫܚܛ
274. ܫܕܐ
275. ܫܕܪ
276. ܫܘܐ
277. ܫܛ
278. ܫܝܛ
279. ܫܢܝ
280. ܫܟܩ
281. ܫܟܚ
282. ܫܠܐ
283. ܫܠܚ
284. ܫܠܛ
285. ܫܠܡ
286. ܫܦܠܗ
287. ܫܦܠ
288. ܫܡܫ
289. ܫܢܐ
290. ܫܢܩ
291. ܫܓܐ
292. ܫܦܪ
293. ܫܩܐ
294. ܫܩܠ
295. ܫܪܐ
296. ܫܪ
297 ܫܬܐ
298. ܫܘܬܦ
299. ܫܬܩ
300. ܬܒܠ

301.	ܬܘܒ	303.	ܬܡܠ	305.	ܬܡܘܗ	307.	ܬܩܢ
302.	ܬܒ	304.	ܬܢܐ	306.	ܬܡܠ	308.	ܬܪܝ

LIST II.

Nouns occurring ten times or more.

1.	ܐܒܐ	23.	ܐܢܫ	45.	ܒܪܝܬܐ	67.	ܕܗܒܐ
2.	ܐܒܕܢܐ	24.	ܐܢܬܬܐ	46.	ܒܠܐ	68.	ܕܘܟ
3.	ܐܓܪܐ	25.	ܐܣܘܪܐ	47.	ܒܥܠܕܒܒܐ	69.	ܕܝܠܐ
4.	ܐܓܪܐ	26.	ܐܣܟܝܡܐ	48.	ܒܪ	70	ܕܝܢܐ
5.	ܐܓܪܬܐ	27.	ܐܦ	49.	ܒܪܘܢ	71.	ܕܡܐ
6.	ܐܕܢܐ	28.	ܐܦܝܢ	50.	ܒܪܝܬܐ	72.	ܕܡܟ
7.	ܐܘܚܕܢܐ	29.	ܐܪܓܘܢܐ	51.	ܒܪܬܐ	73.	ܕܝܢ
8.	ܐܘܡܬܐ	30.	ܐܪܥܐ	52.	ܒܪܫ	74.	ܕܡܬܐ
9.	ܐܘܪܚܐ	31.	ܐܪܡܠܐ	53.	ܒܬܘܠܐ	75.	ܕܡܘܬܐ
10.	ܐܚܐ	32.	ܐܪܡܠܬܐ	54.	ܒܬܪ	76.	ܕܡܪ
11.	ܐܚܪܝܢ	33.	ܐܪܙܐ	55.	ܓܒܪܐ	77.	ܕܝܨܝܬ
12.	ܐܚܪܐ	34.	ܐܬܐ	56.	ܓܕܝܕܐ	78.	ܕܥܐ
13.	ܐܚܝܢܐ	35.	ܐܬܪ	57.	ܓܘ	79.	ܕܚܠܬܐ
14.	ܐܚܝܕܘܬܐ	36.	ܒܗܬܬܐ	58.	ܓܘܪܬܐ	80.	ܕܪܐ
15.	ܐܛܠܡܘܪܐ	37.	ܒܝܬܐ	59.	ܓܘܫܡܐ	81.	ܕܪܓ
16.	ܐܝܩܢܐ	38.	ܒܝܬ	60.	ܓܙܪ	82.	ܕܪܟܬܐ
17.	ܐܠܗܐ	39.	ܒܝܬܗ	61.	ܓܠܝܢܐ	83.	ܕܪܫ
18.	ܐܠܦܐ	40.	ܒܣܪܐ	62.	ܓܢܒ	84.	ܕܪܐ
19.	ܐܠܦܐ	41.	ܒܥܕܢܐ	63.	ܓܢܬܐ	85.	ܗܐ
20.	ܐܡܐ	42.	ܒܨܝܪ	64.	ܓܪܒܐ	86.	ܗܝܡܢܘܬܐ
21.	ܐܡܝܢ	43.	ܒܪ	65.	ܕܒܚܐ	87.	ܗܕܪ
22.	ܐܡܘܣ	44.	ܒܣܪ	66.	ܕܓܠܐ	88.	ܗܘܦܟܐ

89. ܗܦܟ
90. ܗܝܟܠܐ
91. ܗܝܡܢܘܬܐ
92. ܗܕܡܐ
93. ܗܘ
94. ܘܗܒ
95. ܙܕܝܩܐ
96. ܙܒܢܐ
97. ܙܕܝܩܘܬܐ
98. ܙܪܥܐ
99. ܙܘܓܐ
100. ܙܡܪܐ
101. ܙܥܘܪܐ
102. ܙܟܘܬܐ
103. ܙܥܘܪ
104. ܙܩܝܦܐ
105. ܙܝܢܐ
106. ܚܐܪܐ
107. ܚܟܝܡܐ
108. ܚܝܠܐ
109. ܚܡܪܐ
110. ܚܙܐ
111. ܚܕܘܬܐ
112. ܚܘܪ
113. ܚܙܐ
114. ܚܕܥܐ
115. ܚܕܢܐ
116. ܚܕܬܐ
117. ܚܕܪܐ
118. ܚܘܒܐ
119. ܚܘܪܐ
120. ܚܙܝܪܐ
121. ܚܛܗ
122. ܚܛܝܬܐ
123. ܚܛܗܐ
124. ܚܛܝܬܐ
125. ܚܝ
126. ܚܝܘܬܐ
127. ܚܝܐ
128. ܚܝܝܢ
129. ܚܝܠܐ
130. ܚܟܝܡ
131. ܚܟܡܬܐ
132. ܚܠܝܡ
133. ܚܠܦ
134. ܚܠܝܡܐ
135. ܚܡܪܐ
136. ܚܡܫܐ
137. ܚܢܢܐ
138. ܚܢܦܐ
139. ܚܣܕ
140. ܚܣܝܪܘܬܐ
141. ܚܫܟܐ
142. ܚܫܝܫܘ
143. ܚܩܠܐ
144. ܚܪܒ
145. ܚܪܬܐ
146. ܚܫܐ
147. ܛܒܘܬܐ
148. ܛܠܝܘܬܐ
149. ܛܠܐ
150. ܛܠܠܐ
151. ܝܗܒ
152. ܝܡܐ
153. ܝܘܡܐ
154. ܝܘܠܦܢܐ
155. ܝܠܕܬ
156. ܝܕܥ
157. ܝܩܕ
158. ܝܠܐ
159. ܝܡܝܢܐ
160. ܝܢܩܐ
161. ܝܩܝܪܘܬܐ
162. ܝܬܝܪܘܬܐ
163. ܟܐܦܐ
164. ܟܒܪܐ
165. ܟܗܢܘܬܐ
166. ܟܘܟܒܐ
167. ܟܘܠ
168. ܟܘܡܪܐ
169. ܟܘܪܣܝܐ
170. ܟܝܢܐ
171. ܟܐܒܐ
172. ܟܐܒܘܪܐ
173. ܟܢܫ
174. ܟܦܪ
175. ܟܪܣܛܝܢܘܬܐ
176. ܟܬܒ
177. ܟܬܒܢܝܬܐ
178. ܠܒܐ
179. ܠܒܫ
180. ܠܒܘܫܐ
181. ܠܚܡ
182. ܠܝ
183. ܠܠܝܐ
184. ܡܕܡ
185. ܡܕܒܪܐ
186. ܡܕܝܢܬܐ
187. ܡܕܡ
188. ܡܗܝܡܢ
189. ܡܗܝܪܐ
190. ܡܘܬ
191. ܡܚܝܠܐ
192. ܡܠܟܘܬܐ
193. ܡܠܟܐ
194. ܡܢܐ
195. ܡܢܐ
196. ܡܣܟܢ
197. ܡܫܝܚܐ
198. ܡܪܝܐ
199. ܡܪܝܐ
200. ܡܫܚܐ
201. ܢܒܝܐ
202. ܢܗܘܪܐ
203. ܢܘܪܐ
204. ܢܘܟܪܝܐ

205. ܠܠܝܐ
206. ܠܫܢܐ
207. ܡܐܟܘܠܬܐ
208. ܡܐܬܝܬܐ
209. ܡܓܢ
210. ܡܓܪܐ ܕܘܫܐ
211. ܡܕܒܚܐ
212. ܡܕܒܪܐ
213. ܡܕܝܢܬܐ
214. ܡܕܡ
215. ܡܕܝ
216. ܡܘܗܒܬܐ
217. ܡܘܡ
218. ܡܘܬܐ
219. ܡܚܬܐ
220. ܡܚܣܢܐ
221. ܡܚܫܒܬܐ
222. ܡܛܪܐ
223. ܡܝܐ
224. ܡܟܘܪܐ
225. ܡܠܟܬܐ
226. ܡܠܟܘܬܐ
227. ܡܠܚܐ
228. ܡܠܐܟܐ
229. ܡܠܟܐ
230. ܡܠܟܐ
231. ܡܠܟܢܐ
232. ܡܠܬܐ
233. ܡܢܝܢܐ
234. ܡܢܬܐ
235. ܡܣܒ ܒܐܦܐ
236. ܡܣܝܒܪܢܘܬܐ
237. ܡܣܟܝܢܐ
238. ܡܣܢܐ
239. ܡܥܕܪܢܐ
240. ܡܥܝܢܐ
241. ܡܥܡܘܕܝܬܐ
242. ܡܦܩ
243. ܡܨܥܬܐ
244. ܡܪ
245. ܡܪܐ
246. ܡܫܚܐ
247. ܡܫܟܢܐ
248. ܡܫܡܫܢܐ
249. ܡܫܡܫܘܬܐ
250. ܡܬܠܐ
251. ܢܒܝܐ
252. ܢܒܝܘܬܐ
253. ܢܗܝܪܐ
254. ܢܗܪܐ
255. ܢܘܟܪܝܐ
256. ܢܘܢܐ
257. ܢܘܪܐ
258. ܢܚܬܐ
259. ܢܟܣܐ
260. ܢܟܠܐ
261. ܢܡܘܣܐ
262. ܢܨܚܢܐ
263. ܣܒܐ
264. ܣܓܝܐܐ
265. ܣܒܪܐ
266. ܣܒܪܬܐ
267. ܣܓܕ
268. ܣܗܕܐ
269. ܣܘܦܐ
270. ܣܘܥܪܢܐ
271. ܣܛܢܐ
272. ܣܝܒܪܬܐ
273. ܣܝܡܬܐ
274. ܣܝܦܐ
275. ܣܟܠܐ
276. ܣܠܩ
277. ܣܡܐ
278. ܣܡܝܐ
279. ܣܦܪܐ
280. ܣܦܝܢܬܐ
281. ܣܦܩ
282. ܣܪܗܒ
283. ܥܒܕ
284. ܥܒܕ
285. ܥܒܕܘܬܐ
286. ܥܒܪܐ
287. ܥܓܠ
288. ܥܕܡܐ
289. ܥܕܥܕܐ
290. ܥܕܬܐ
291. ܥܘܠܐ
292. ܥܘܠܐ
293. ܥܘܪܘܬܐ
294. ܥܘܬܪܐ
295. ܥܙܐ
296. ܥܝܢ
297. ܥܠ
298. ܥܠܡܐ
299. ܥܠܡ
300. ܥܡ
301. ܥܡܠܐ
302. ܥܢܐ
303. ܥܢܢܐ
304. ܥܦܪܐ
305. ܥܪܣܐ
306. ܥܪܩ
307. ܥܪܒܐ
308. ܥܬܝܕ
309. ܥܬܝܩܐ
310. ܥܬܝܪ
311. ܦܐܪܐ
312. ܦܓܪܐ
313. ܦܘܠܚܢܐ
314. ܦܘܡ
315. ܦܘܩܕܢܐ
316. ܦܘܪܥܢܐ
317. ܦܝܠܛܘܣ
318. ܦܠܓܘܬܐ
319. ܦܪܘܩܐ
320. ܦܪܨܘܦܐ

321. ܦܘܫܟܐ
322. ܦܘܫܪܐ
323. ܦܘܘܪܐ
324. ܦܨܗ
325. ܦܨܢܬܐ
326. ܦܪܡ
327. ܦܟܗ
328. ܦܟܨܟܐ
329. ܦܢܟܗܐ
330. ܦܩܪܐ
331. ܨܨܕܪܐ
332. ܨܨܪܐ
333. ܨܡܡܐ
334. ܨܡܕܠܐ
335. ܨܡܨܕ
336. ܨܠܘܬܐ
337. ܨܠܘܐ
338. ܨܕܡܗܐ
339. ܨܝܡܪܐ
340. ܨܡܘܡܐ
341. ܨܡܥܬܐ
342. ܨܡܚܐ
343. ܨܢܐ
344. ܨܒܠܐ
345. ܨܦܪܐ
346. ܩܢܝܐ
347. ܩܢܘܢܐ
348. ܩܪܐ
349. ܩܪܝܒܐ
350. ܩܪܝܢܐ
351. ܩܪܢܐ
352. ܩܫܝܫܐ
353. ܪܐܙܐ
354. ܪܒܐ
355. ܪܒܘܬܐ
356. ܪܒܗ
357. ܪܓܠܐ
358. ܪܓܠܘܢ
359. ܪܓܝܓ
360. ܪܓܫ
361. ܪܘܚܐ
362. ܪܘܡܐ
363. ܪܚܡܬܐ
364. ܪܚܡܢ
365. ܪܚܡ
366. ܪܚܝܡܗ
367. ܪܟܒܐ
368. ܪܟܝܟܐ
369. ܪܥܝܐ
370. ܪܥܝܢܐ
371. ܪܫܝܢ
372. ܪܫܡܬܐ
373. ܫܒܝܐ
374. ܫܒܬܐ
375. ܫܓܝܫܘܬܐ
376. ܫܘܒܗܪܐ
377. ܫܘܒܚܐ
378. ܫܘܘܕܝܐ
379. ܫܘܠܛܢ
380. ܫܘܠܛܢܐ
381. ܫܘܠܡܐ
382. ܫܘܩܐ
383. ܫܘܬܦܐ
384. ܫܘܬܦܘܬܐ
385. ܫܚܘܪ
386. ܫܠܝܐ
387. ܫܠܡܐ
388. ܫܠܝܚܐ
389. ܫܠܝܛܐ
390. ܫܠܝܡ
391. ܫܡܝܐ
392. ܫܡܥܢܐ
393. ܫܡܫܐ
394. ܫܢܐ
395. ܫܢܬܐ
396. ܫܦܝܪ
397. ܫܪܟܐ
398. ܫܪܝܪܐ
399. ܫܪܥܐ
400. ܫܪܪܐ
401. ܫܬܩܬܐ
402. ܫܬܐܣܬܐ
403. ܬܐܪܬܐ
404. ܬܘܒ
405. ܬܘܕܝܬܐ
406. ܬܘܟܠܢܐ
407. ܬܚܘܡܐ
408. ܬܡܢ
409. ܬܢܢܐ
410. ܬܪܢܓܠܐ
411. ܬܪܘܨ
412. ܬܪܥܐ
413. ܬܪܝܨܘܬܐ
414. ܬܫܒܚܬܐ
415. ܬܫܡܫܬܐ
416. ܓܒܐ
417. ܨܠܝ
418. ܩܘܠܠܢܐ
419. ܩܘܡܘܡܬܐ

WORD LISTS—ENGLISH.

LIST I.

Verbs occurring ten times or more in Schaaf's Concordance of the New Testament.

1. perish
2. mourn
3. trade
4. go
5. seize
6. delay
7. there is
8. eat
9. learn
10. compel
11. believe
12. say
13. heal
14. bind
15. meet
16. pour
17. come
18. be evil
19. scathe
20. be ashamed
21. laugh at
22. cease
23. conceive
24. console
25. weep
26. devour
27. build
28. despise
29. be sweet
30. cry
31. search
32. decrease
33. prove
34. create
35. bless
36. choose
37. commit adultery
38. circumcise
39. reveal
40. complete
41. steal
42. commend
43. sacrifice
44. lead
45. lie
46. judge
47. trample
48. fear
49. be pure
50. remember
51. disturb
52. be like
53. sleep
54. wonder
55. rise
56. quench
57. tie
58. seek
59. be
60. go
61. turn
62. injure
63. it is right
64. bring
65. be just
66. shine
67. be moved
68. conquer

69. sing
70. be a harlot
71. crucify
72. sow
73. love
74. corrupt
75. bind
76. rejoice
77. surround
78. owe
79. show
80. strengthen
81. spare
82. look
83. see
84. bind
85. sin
86. snatch
87. live
88. sleep
89. change
90. be gracious
91. strangle
92. want
93. be diligent
94. reap
95. free
96. think
97. be useful
98. be dark
99. suffer
100. seal
101. be proud
102. defraud
103. envy
104. err
105. taste
106. bear
107. obey
108. hide
109. sound
110. lead
111. to be dry
112. thank
113. know
114. give
115. bow
116. learn
117. swear
118. add
119. care
120. burn
121. honor
122. be great
123. inherit
124. extend
125. sit
126. abound
127. rebuke
128. correct
129. prohibit
130. collect
131. conceal
132. accuse
133. hunger
134. deny
135. to be sad
136. preach
137. wrap
138. offend
139. write
140. continue
141. fight
142. be weary
143. encourage
144. clothe
145. accompany
146. curse
147. there is not
148. eat
149. be grieved
150. die
151. smite
152. come
153. be humble
154. be full
155. counsel
156. speak
157. be able
158. dare
159. anoint
160. prophesy
161. strike
162. abide
163. shine
164. rest
165. go down
166. guard
167. kill
168. be sober
169. receive
170. tempt
171. ascend
172. fall
173. go out
174. plant
175. cleave to
176. beat
177. kiss
178. give
179. defile
180. satisfy
181. hope
182. be great
183. worship
184. witness

185. put
186. subvert
187. expect
188. perceive
189. reject
190. ascend
191. lie down
192. hate
193. be poor
194. work
195. suffice
196. be vacant
197. destroy
198. make
199. pass over
200. aid
201. remember
202. grieve
203. watch
204. cover
205. enter
206. baptize
207. labor
208. inhabit
209. answer
210. flee
211. to be rich
212. persuade
213. divide
214. serve
215. return
216. permit
217. decree
218. do
219. liberate
220. command
221. be tolerable
222. fly
223. remunerate
224. depart
225. separate
226. extend
227. expound
228. open
229. wish
230. dip
231. thirst
232. hunt
233. fast
234. revile
235. pray
236. be vile
237. lacerate
238. receive
239. bury
240. sanctify
241. remain
242. arise
243. kill
244. possess
245. cry
246. break
247. call
248. be nigh
249. be hard
250. be great
251. desire
252. be angry
253. stone
254. go
255. pursue
256. disturb
257. run
258. be inebriated
259. exult
260. be high
261. love
262. murmur
263. ride
264. cast
265. meditate
266. feed
267. lament
268. ask
269. take captive
270. be glorious
271. praise
272. lean
273. be tumultuous
274. hurl
275. send
276. be worthy
277. wash
278. despise
279. compel
280. sleep
281. be able
282. be at rest
283. send
284. rule
285. finish
286. name
287. hear
288. minister
289. change
290. torment
291. narrate
292. please
293. drink
294. bear
295. loose
296. confirm
297. drink
298. communicate
299. be silent
300. arrange

301. repent
302. return
303. hope
304. hang
305. admire
306. offend
307. prepare
308. direct

LIST II.

Nouns and other words occurring ten times or more.

1. father
2. loss
3. hire
4. hireling
5. letter
6. ear
7. affliction
8. water
9. way
10. brother
11. other
12. hand
13. tree
14. day
15. calumniator
16. stranger
17. god
18. thousand
19. ship
20. mother
21. Amen
22. when
23. man
24. woman
25. chain
26. scheme
27. also
28. face
29. purple
30. lion
31. Gentile
32. widow
33. earth
34. sign
35. place
36. shame
37. consolation
38. bad
39. evil
40. house
41. building
42. sweet
43. flesh
44. behind
45. petition
46. end
47. evening
48. son
49. knee
50. creature
51. creator
52. but
53. origin
54. after
55. man
56. gehenna
57. midst
58. circumcision
59. robber
60. for
61. revelation
62. side
63. thief
64. leper
65. sacrifice
66. liar
67. gold
68. place
69. fearful
70. fear
71. demon
72. own
73. but
74. judgment
75. judge
76. denar
77. covenant
78. pure
79. purity
80. male
81. blood
82. likeness
83. tear
84. generation
85. behold
86. governor
87. member
88. overthrow

89. then
90. temple
91. faith
92. here
93. woe
94. time
95. Sadducees
96. righteous
97. righteousness
98. alms
99. movement
100. olive
101. victory
102. adultery
103. little
104. cross
105. seed
106. free
107. beloved
108. corruption
109. neighbor
110. one
111. joy
112. around
113. new
114. love
115. serpent
116. rod
117. white
118. desolation
119. vision
120. swine
121. sin
122. sin
123. sinner
124. wheat
125. living
126. debtor
127. animal
128. life
129. strength
130. wise
131. wisdom
132. sound
133. in exchange for
134. ferment
135. wine
136. wrath
137. grace
138. profane
139. wanting
140. want
141. zeal
142. perseverance
143. back
144. mute
145. end
146. suffering
147. darkness
148. supper
149. sister
150. husband
151. good
152. report
153. happiness
154. family
155. error
156. rock
157. goodness
158. youth
159. unclean
160. impure
161. impurity
162. error
163. beautiful
164. knowledge
165. Jew
166. teaching
167. day
168. to-day
169. gain
170. glory
171. child
172. begetter
173. glorious
174. month
175. inheritance
176. more
177. more
178. sorrow
179. just
180. justice
181. already
182. when
183. priest
184. star
185. priest
186. sickness
187. throne
188. nature
189. stone
190. all
191. crown
192. synagogue
193. crown
194. fellow
195. cup
196. silver
197. sick
198. belly
199. vineyard
200. book
201. heart
202. clothing
203. alone
204. bread

205. night
206. tongue
207. food
208. advent
209. for nought
210. patience
211. last
212. desert
213. city
214. anything
215. knowledge
216. gift
217. spot
218. death
219. stroke
220. saviour
221. thought
222. rain
223. water
224. excellent
225. humble
226. humility
227. publican
228. angel
229. king
230. counsel
231. kingdom
232. word
233. number
234. part
235. hypocrisy
236. endurance
237. poor
238. shoe
239. helper
240. fountain
241. baptism
242. defense
243. midst
244. lord
245. Lord
246. ointment
247. tabernacle
248. traitor
249. banquet
250. parable
251. prophet
252. prophecy
253. light
254. river
255. strange
256. fish
257. fire
258. vestment
259. rest
260. guile
261. law
262. temptation
263. soul
264. splendid
265. hope
266. gospel
267. much
268. witness
269. branch
270. work
271. Satan
272. food
273. treasure
274. sword
275. food
276. foolishness
277. blind
278. reclining
279. hair
280. ship
281. scribe
282. vain
283. servant
284. service
285. servitude
286. further
287. quickly
288. time
289. feast
290. church
291. iniquity
292. wicked
293. foreskin
294. wishes
295. custom
296. eye
297. cause
298. youth
299. ever
300. people
301. labor
302. flock
303. cloud
304. root
305. sheep
306. naked
307. bed
308. future
309. old
310. rich
311. fruit
312. body
313. work
314. mouth
315. command
316. redemption
317. phial
318. division
319. saviour
320. face

321. word
322. idol
323. table
324. thing
325. will
326. with
327. prayer
328. image
329. snare
330. morning
331. sepulture
332. sepulchre
333. holy
334. holiness
335. first
336. gift
337. field
338. truth
339. force
340. resurrection
341. stable
342. word
343. voice
344. little
345. self
346. reed
347. possession
348. city
349. war
350. call
351. hour
352. elder
353. secret
354. great
355. magnitude
356. myriad
357. desire
358. wrath
359. desirable
360. foot
361. spirit
362. afar
363. far
364. mercy
365. head
366. beginning
367. high
368. evening
369. mind
370. thunder
371. blame
372. impious
373. sceptre
374. sabbath
375. tumult
376. glory
377. praise
378. promise
379. rule
380. end
381. rock
382. market-place
383. partaker
384. communion
385. sheol
386. peace
387. rest
388. apostle
389. ruler
390. peace
391. name
392. heaven
393. sun
394. year
395. hour
396. good
397. family
398. true
399. remainder
400. truth
401. chain
402. foundation
403. conscience
404. again
405. thanks
406. confidence
407. disciple
408. then
409. smoke
410. cock
411. throne
412. gate
413. mind
414. glory
415. service
416. fig-tree
417. vessel
418. promise
419. oath

TRANSLITERATION OF GENESIS I.

1. B^{e}ri-shith b^{e}ro’ ’aloho’ yoth sh^{e}mayo’ w^{e}yoth ’arʿo’.

2. Wa’rʿo’ h^{e}woth tuh w^{e}bhuh w^{e}ḥeshshukho’ ʿal ’appay t^{e}humo’ w^{e}ruḥeh da’loho’ m^{e}raḥḥ^{e}pho’ ʿal ’appay mayo’ we’mar ’aloho’ nehwe’ nuhro’ wah^{e}wo’ nuhro’.

3. Waḥ^{e}zo’ ’aloho’ l^{e}nuhro’ dh^{e}shappir.

4. Waph^{e}rash ’aloho’ bhêth nuhro’ l^{e}ḥeshshukho’.

5. Waḳ^{e}ro’ ’aloho’ l^{e}nuhro’ ’îmomo’ wal^{e}ḥeshshukho’ ḳ^{e}ro’ lelyo’ wah^{e}wo’ ramsho’ wah^{e}wo’ ṣaphro’ yaumo’ ḥadh.

6. We’mar’ aloho’ nehwe’ r^{e}ḳiʿo’ bh^{e}metsʿath mayo’ w^{e}nehwe’ phoresh bêth mayo’ l^{e}mayo’.

7. Waʿ^{e}bhadh ’aloho’ ’arḳiʿo’ waph^{e}rash bêth mayo’ dal^{e}thaḥt men ’arḳiʿo’ w^{e}bhêth mayo’ dal^{e}ʿel men ’arḳiʿo’ wah^{e}wo’ hokhanno’.

8. Waḳ^{e}ro’ ’aloho’ la’rḳiʿo’ sh^{e}mayo’ wah^{e}wo’ ramsho’ wah^{e}wo’ ṣaphro’ yaumo’ dhath^{e}rên.

9. We’mar ’aloho’ nethkann^{e}shun mayo’ dhal^{e}thaḥt men sh^{e}mayo’ la’thro’ ḥadh w^{e}thethḥ^{e}ze’ yabbishto’ wah^{e}wo’ hokhanno’.

10. Waḳ^{e}ro’ ’aloho’ l^{e}yabbishto’ ’arʿo’ wal^{e}khensho’ dh^{e}mayo’ ḳ^{e}ro’ ya(m)me’ waḥ^{e}zo’ ’aloho’ dh^{e}shappir.

11. We’mar ’aloho’ thappeḳ ’arʿo’ thadho” ʿesbo’ dh^{e}mezd^{e}raʿ zarʿo’ l^{e}ghenseh wi’ylono’ dh^{e}phi’re’ dh^{e}ʿobhedh pi’re’ l^{e}ghenseh d^{e}neṣb^{e}theh beh ʿal ’arʿo’ wah^{e}wo’ hokhanno’.

12. Wapp^{e}ḳath ’arʿo’ thadho” ʿesbo’ dh^{e}mezd^{e}raʿ zarʿo’ l^{e}ghenseh wi’ylono dh^{e}ʿobhedh pi’re’ dh^{e}neṣb^{e}theh beh l^{e}ghenseh waḥ^{e}zo’ ’aloho’ dh^{e}shappir.

13. Wah^{e}wo’ ramsho’ wah^{e}wo’ ṣaphro’ yaumo’ dhath^{e}lotho’.

14. We'mar 'aloho' nehwun nahhîre' ba'rkîʿo' dhash^{e}mayo' l^{e}mephrash bêth îmomo' l^{e}lelyo' w^{e}nehwun lo'th^{e}wotho' wal^{e}zabhnê' wal^{e}yaumotho' w^{e}lash^{e}nayo'.

15. W^{e}nehwun manh^{e}rîn ba'rkîʿo' dhash^{e}mayo' l^{e}manhoru ʿal 'arʿo' wah^{e}wo' hokhanno'.

16. Waʿ^{e}bhadh 'aloho' th^{e}rên nahhîrê' raur^{e}bhê' nahhîro' rabbo' l^{e}shulṭono dhi'ymomo' w^{e}nahhîro' z^{e}ʿuro' l^{e}shulṭono' dh^{e}lelyo' w^{e}khauk^{e}bhê'.

17. W^{e}yahbh 'ennun 'aloho' bha'rḳîʿo' dhash^{e}mayo' l^{e}manhoru ʿal 'arʿo'.

18. Wal^{e}meshlaṭ bî'ymomo' wabh^{e}lelyo' wal^{e}mephrash bêth nuhro' l^{e}ḥeshshukho' waḥ^{e}zo' 'aloho' dh^{e}shappir.

19. Wah^{e}wo' ramsho' wah^{e}wo' ṣaphro' yaumo' dha'rb^{e}ʿo'.

20. We'mar 'aloho' narḥ^{e}shun mayo' raḥsho' napsho' ḥayy^{e}tho' w^{e}phoraḥ^{e}tho' thephraḥy ʿal 'arʿo' ʿal 'appay 'arkîʿo' dhash^{e}mayo'.

21. Wabh^{e}ro' 'aloho' thannîne' raur^{e}bhe' w^{e}khul naphsho' ḥayy^{e}tho' dh^{e}raḥsho' dha'rḥeshw mayo' l^{e}ghens^{e}hun w^{e}khul poraḥ^{e}tho' dh^{e}gheppo' l^{e}ghensoh waḥ^{e}zo' 'aloho' dh^{e}shappir.

22. W^{e}bharrekh 'ennun 'aloho' we'mar l^{e}hun, p^{e}rau was^{e}ghau wam^{e}lau mayo' dhabh^{e}ya(m)me'. w^{e}phoraḥ^{e}tho' thesge' bha'rʿo'.

23. Wah^{e}wo' ramsho' wah^{e}wo' ṣaphro' yaumo' dh^{e}ḥamsho'.

24. We'mar 'aloho' thappeḳy 'arʿo' naphsho' ḥayy^{e}tho' l^{e}ghensoh b^{e}ʿîro' w^{e}raḥsho' w^{e}hayw^{e}tho'. dha'rʿo' l^{e}ghensoh wah^{e}wo' hokhanno'.

25. Waʿ^{e}bhadh 'aloho' hayw^{e}tho' dha'rʿo' l^{e}ghensoh wabh^{e}ʿîro' l^{e}ghensoh w^{e}khulleh raḥsho' dha'rʿo' l^{e}ghensauhy wah^{e}zo' 'aloho' dh^{e}shappir.

26. We'mar 'aloho' neʿbedh 'nosho' bh^{e}ṣalman 'aykh d^{e}muthan w^{e}neshl^{e}ṭun b^{e}nunay yammo' wabh^{e}phoraḥ^{e}tho' dhash^{e}mayo' w^{e}bhabh^{e}ʿîro' wabh^{e}khulloh hayw^{e}tho' dha'rʿo' wabh^{e}khulloh raḥsho' dh^{e}roḥesh ʿal 'arʿo'.

27. Wabh^{e}ro' 'aloho' lo'dhom b^{e}ṣalmeh baṣ^{e}lem 'aloho' b^{e}royhy d^{e}khar w^{e}neḳbo' bh^{e}ro' 'ennun.

28. W^{e}bharrekh ’ennun ’aloho’ we’mar l^{e}hun ’aloho’ ph^{e}rau was^{e}ghau wam^{e}lau ’arʿo’ w^{e}khubhshuh wash^{e}laṭw b^{e}nunay-yammo’. wabh^{e}phoraḥ^{e}tho’ dhash^{e}mayo’ w^{e}bhabh^{eʿ}îro’ wabh^{e}khulloh hayw^{e}tho’ dh^{e}roḥsho’ ʿal ’arʿo’.

29. We’mar ’aloho’ ho’ yehbeth l^{e}khun kulleh ʿesbo’ dh^{e}zarʿo’ dh^{e}mezd^{e}raʿ ʿal ’appay kulloh ’arʿo’ w^{e}khul ’îlon di’yth beh pi’ray ’îloneh d^{e}zarʿeh mezd^{e}raʿ l^{e}khun nehwe’ me’khulto’ wal^{e}khulloh hayw^{e}tho’ dh^{e}dhabhro’.

30. Wal^{e}khulloh poraḥ^{e}tho’ dh’ash^{e}mayo’ wal^{e}khul d^{e}roḥesh ʿal ’arʿo’ dhi’yth beh naphsho’ hayy^{e}tho’ w^{e}khulleh yurroḳo’ dh^{eʿ}esbo’ l^{e}me’khulto’ wah^{e}wo’ hokhanno’.

31. Waḥ^{e}zo’ ’aloho’ khul daʿ^{e}badh w^{e}ho’ ṭobh shappîr wah^{e}wo’ ramsho’ wah^{e}wo’ ṣaphro’ yaumo’ dheshto’.

GENESIS I.–IV.

A LITERAL TRANSLATION.*

CHAPTER I.

1. In beginning created the God + the heavens and + the earth.

2. And the earth was tuh and buh and the darkness (was) upon the faces of the abyss, and the spirit of him who (is) the God (was) brooding upon the faces of the waters, and said the God, let be the light, and was the light.

3. And saw the God + the light that (it was) good.

4. And separated the God between the light to the darkness.

5. And called the God to the light the day and to the darkness called he the night and it was the evening and it was the morning the day one.

6. And said the God let be the expanse in the midst of the waters and let it be separating between the waters to the waters.

7. And made the God the expanse and separated between the waters which to under from the expanse and between the waters which to above from the expanse, and it was so.

8. And called the God to the expanse the heavens, and it was the evening and it was the morning the day which (is) two.

9. And said the God: let be assembled the waters which (are) to under from the heavens to the place one and let be seen the dry land, and it was so.

10. And called the God to the dry land the earth and to the gathering of the waters called he the seas, and saw the God that (it was) good.

---

* The + sign denotes some particle in Syriac which cannot be translated into English. Words in parenthesis occur in English but not in Syriac.

11. And said the God ; Let cause to go out the earth the grass the herb which is seeding for itself the seed (according) to the kind his and the tree that of the fruits which (is) making the fruits (according) to kind his which sprout his (is) in him upon the earth, and it was so.

12. And caused to go out the earth the grass, the herb which (is) seeding for itself the seed (according) to kind his and the tree which (is) making the fruits which sprout his (is) in him (according) to kind his, and saw the God that it was good.

13. And it was the evening and it was the morning the day which (is) three.

14. And said the God let be the light in the expanse that of the heavens to separate between the day to the night, and let them be for the signs and for the times and for the days and for the years.

15. And let them be giving light in the expanse that of the heavens to give light upon the earth, and it was so.

16. And made the God two the lights the great, the light the great for the ruling that of the day and the light the less for the ruling that of the night, and the stars.

17. And gave them the God in the expanse that of the heavens to give light upon the earth.

18. And to rule in the day and in the night and to separate between the light to the darkness, and saw the God that (it was) good.

19. And it was the evening and it was the morning the day which (is) four.

20. And said the God : Let swarm the waters the swarm the soul the living and the bird let her fly upon the earth upon (the) faces of the expanse that of the heavens.

21. And created the God the sea-monsters the great and every (one) the soul the living that of the swarm which swarmed the waters (according) to kind their and every one the bird the living (according) to kind her and saw the God that (it was) good.

22. And blessed them the God and said to them: Be fruitful and multiply and fill the waters which are in the seas and the bird let it multiply in the earth.

23. And it was the evening and it was the morning the day which (is) five.

24. And said the God: Let cause to go out the earth the soul the living (according) to kind her, the cattle and swarm and the animal that of the earth (according) to kind her and all of him the swarm that of the earth according to kind her, and it was so.

25. And made the God the animal that of the earth (according) to kind her and the cattle according to kind her and all of him the swarm that of the earth according to kind his and saw the God that (it was) good.

26. And said the God: Let us make the man in image our according to likeness our and let them rule over the fish of the sea and over the bird that of the heavens and over the cattle and over all of her the animal that of the earth and over all of them the swarm which is swarming upon the earth.

27. And created the God + man in image his in (the) image of the God created he him, male and female created he them.

28. And blessed them the God and said to them: Be fruitful and multiply and fill the earth and subdue her, and rule over the fish of the sea and over the bird that of the heavens and over the cattle and over all of the animal which is swarming upon the earth.

29. And said the God: Behold I have given to you all of him the herb that of seed which is seeding for itself upon (the) faces of all of her the earth and every tree which exists in him (the) fruits of tree his which seed his (is) seeding for himself. To you shall it be the food and to all of her the animal that of the field.

30. And to all of her the bird that of the heavens and to all which swarmeth upon the earth which exists in it the soul the living and all of him the green that of the herb (shall be) for the food, and it was so.

31. And saw the God all which he had made and behold (it was) very good and it was the evening and it was the morning the day which is six.

CHAPTER II.

1. And were finished the heavens and the earth and all [of him] their host.
2. And finished the God in the day the sixth works his which he had made and he rested himself in the day the seventh from all of them, works his, which he made.
3. And blessed the God + the day the seventh and sanctified him because that in him he had rested himself from all of them, works his, which created the God by making.
4. These (are) the generations those of the heavens and those of the earth when they were created in the day (in) which made the Lord the God the heavens and the earth.
5. And all of them, the trees those of the field, as yet not had been in the earth and all of him the herb that of the field as yet not had gone out, because that not had caused to come down the Lord the God the rain upon (the) faces of the earth and Adam existed not to till (in) the earth.
6. And the mist going up had been from the earth and watering had been + all (the) faces of the earth.
7. And formed the Lord the God + Adam (of) the dust from the ground and breathed into nostrils his the breath that of the lives and was Adam to (a) soul (a) living.
8. And planted the Lord the God the Paradise in Eden from east and put there + Adam whom he had formed.
9. And caused to go out the Lord the God from the earth every tree which (is) pleasant to see and good to eat and the tree of the lives in the midst of him that is the park and the tree that of the knowledge that of the good and that of the evil.
10. And the river going was from Eden to water him + the park and from there (it was) separating and becoming four heads.
11. The name of him that (is) one (is) Pishun; he (is) that surrounding + all of her the land that of H^{e}wilo which there (is) gold.

12. And the gold of her that (is) the land, that (is) good; there (are) bdellium and the stones which (are) the beryl.

13. And the name of him that of the river the second (is) Gishun, that (is) that which (is) surrounding+all of her the land that of Kush.

14. And the name of him that of the river which is three (is) Tigris, that (is) that which (is) going before Assyria and the river which is four he (is) Euphrates.

15. And took the Lord the God+Adam and left him in the park that of Eden that he might till him and keep him.

16. And commanded the Lord the God+Adam and said to him: From all of them the trees those which (are) in the park thou mayest eat.

17. And from the tree that of the knowledge that of the good and that of the evil not shalt thou eat from him, because that in the day (in) which thou shalt eat from him the death thou shalt die.

18. And said the Lord the God: Not (is it) good that should be Adam in solitariness his [*i.e.*, alone]. I will make for him the helper like him.

19. And formed the Lord the God from the earth all of her the animal that of the field and all of her the bird that of the heavens and brought them unto Adam that he might see what (he was) calling+them, and all which called to them Adam the soul the living, that *is* his name.

20. And called Adam the names to all of her the cattle and to all of her the bird that of the heavens and to all of her the animal that of the earth; and for Adam not was found for him the helper like him.

21. And cast the Lord the God the rest upon Adam and he slept and he took one from ribs his and closed the flesh instead of her.

22. And constructed the Lord the God the rib which he had taken from Adam into the woman and he brought her to Adam.

23. And said Adam: The this the time the bone (is) from bones of me and the flesh from flesh of me, the this shall be called the woman because that from the man (is she) taken.

24. Because of the this shall leave the man + (the) father of him and + (the) mother of him and shall cleave to (the) wife of him and shall be the two of them one flesh.

25. And they were (the) two of them naked, Adam and the woman of him and not (were they) ashamed.

CHAPTER III.

1. And the serpent was cunning from (*i. e.*, more than) every animal of the field which had made the Lord God and said the serpent to the woman: Truly hath said God that not should ye eat from all the trees of the park?

2. And said the woman to the serpent: (It is true) that from the fruits of the trees which (are) in the park, all of them, we may eat.

3. And from the fruits of the tree which (is) in the midst of him that (is) the park hath said God [that] ye shall not eat from him and ye shall not draw nigh to him lest (*i. e.*, that not) ye die.

4. And said the serpent to the woman: Ye shall not surely die.

5. Because that knows God that in the day that eating (are) ye from him, (shall be) opened your eyes and ye (shall be) existing like God (*i. e.*, as) knowers of the good and the evil.

6. And saw the woman that good (was) the tree for eating and the pleasure he (was) to the eyes and (that) pleasant (was) the tree to look at, and she took from the fruits of him and ate and gave also to her husband with her and he ate.

7. And were opened the eyes those of both of them and they knew that naked (were) they and they sewed the leaves those of the fig-trees and made for them the aprons.

8. And they heard the voice of him who (is) the Lord God (as he was) walking in the park at the turning of him that (is) the day, and they concealed themselves Adam and his wife from before the Lord God in the midst of the trees which (were) in the park.

9. And called the Lord God to Adam and said to him: Where (art) thou Adam?

10. And he said: Thy voice have I heard in the park and I saw that naked (am) I and I hid myself.

11. And said to him the Lord: Who (is) he (that) hath showed thee that naked thou (art)? Behold from the tree (concerning) which I commanded thee that thou shouldest not eat from him thou hast eaten.

12. And said Adam: The woman whom thou gavest (to be) with me she has given to me from the tree and I have eaten. And said the Lord God to the woman.

13. What is this that thou hast done? And said the woman: The serpent deceived me and I ate.

14. And said the Lord God to the serpent: Because thou hast done this, cursed (be) thou above all cattle and above every animal of the field, and upon thy belly shalt thou go and the dust shalt thou eat all of the days of thy lives.

15. And the enmity shall I put between thee to the woman and between thy seed to her seed; *he* shall trample thy head and *thou* shalt smite him in his heel.

16. And to the woman he said: I will surely multiply thy pains and thy conceptions and in pains shalt thou bear sons [children] and unto thy husband shalt thou turn thyself and *he* shall have dominion over thee.

17. And to Adam he said: Because thou hast hearkened to (*lit.*, heard in) the voice of her who is thy wife and hast eaten from the tree (concerning) which I commanded thee and said to thee, that thou shouldest not eat from him, cursed (be) thy land because of thee in the pains shalt thou eat (of) her all of the days of thy lives.

18. Thorns and thistles shall she bring out for thee and thou shalt eat the herb that of the field.

19. And in the sweat that of thy nostrils (or *faces*) shalt thou eat the bread until that thou shalt return to the earth which from her thou hast been taken; because that the dust thou (art) and to the dust thou shalt return.

20. And called Adam the name of her who (was) his wife Eve, because that *she* was the mother that of all which (is) living.

21. And made the Lord God for Adam and for his wife the coats those of the skin and clothed them.

22. And said the Lord God : Behold Adam has become like one of us (as) to the knowing of the good and the evil. Now lest he stretch out his hand and take also from the tree that of the lives and eat and live for ever.

23. And sent him the Lord God from the park that of Eden to till the earth which he was taken from there.

24. And caused him to go out the Lord God and he caused to go around from the east to the park that of Eden the cherub and the flame of the sword which (was) turning itself to keep the way that of the tree that of the lives.

CHAPTER IV.

1. And Adam knew + Eve his wife and she conceived and bare + Cain, and she said : I have gotten the man for the Lord.

2. And she added to bear + his brother Abel. And Abel was feeding the flock and Cain was laboring in (tilling) the earth.

3. And it came to pass after some days that (lit., *and*) Cain brought from the fruits those of his earth (or *ground*) the gift to the Lord.

4. And Abel brought, also he, from the firstlings, those of his flock, and from their fatlings ; and took pleasure the Lord in Abel and in his offering.

5. And in Cain and in his offering he did not take pleasure and it displeased Cain [Note the idiom] exceedingly and was sad his face (*lit.*, were darkened his nostrils or faces).

6. And said the Lord to Cain : Why art thou displeased, and why has become sad thy face ?

7. Behold if thou doest well, I have accepted ; and if not thou doest well, at the door the sin (is) laid, thou wilt turn thyself unto him and *he* shall have dominion over thee.

8. And said Cain to Abel his brother : Let us go to the plain. And it came to pass that when they (were) in the field arose Cain against Abel his brother and killed him.

9. And said the Lord to Cain : Where is Abel thy brother ? And he said : I know not. His keeper (am) I + that of my brother ?

10. And he said: What hast thou done? The voice that of the blood of him who (is) thy brother (is) crying unto me from the earth.

11. Therefore cursed (be) thou from the earth which has opened her mouth and has received the blood of him who (is) thy brother from thy hands.

12. When thou shalt labor in the earth she shall not add that she should give to thee her strength, fleeing and wandering shalt thou be in the earth.

13. And said Cain to the Lord: Great *is* my sin from that which (is) to remit.

14. Behold thou hast caused me to go out to-day from the faces of the earth and from before thee I shall be hidden and I shall be fleeing and wandering in the earth and anyone who shall find me will slay me.

15. And said to him the Lord: Not so, whosoever (is) the killer of Cain, sevenfold shall he be avenged. And put the Lord the sign on Cain that should not kill him every (one) whosoever (should be) finding + him.

16. And went out Cain from before the Lord and he dwelt in the Land that of Nod from east of her which (is) Eden.

17. And knew Cain + his wife and she conceived and bare + Enoch and he built the city and called the name of her which (is) the city after the name of his son Enoch.

18. And was born to Enoch Irad, and Irad begat + Mehuel, and Mehuel begat + Methushel and Methushel begat + Lamech.

19. And Lamech took to him two wives, the name of her that (is) one Adah and the name of her which (is) the second (*lit.*, next, or following) Zillah.

20. And bare Adah + Jobal; *he* was the father to the inhabiters of the tents and the possessors of the possession.

21. And the name that of his brother (was) Jubal; *he* was the father to every (one) who (is) laying hold of the cithara and the kinura.

22. And Zillah also *she* bare + Tubal-Cain, an artificer in every work that of the brass and that of the iron; and the sister of him who (is) Tubal-Cain (was) Naamah.

23. And said Lamech to his wives: Adah and Zillah hear ye my voice, wives of Lamech hearken to my saying; because that the man have I killed by my wounds and the youth by my blow.

24. Because that one in seven (*i. e.*, seven-fold) shall be avenged Cain and Lamech to seventy and seven.

25. And knew Adam again + Eve his wife and she conceived and bare the son and she called his name Seth, because that has given to me God the seed the other instead of Abel that (or because) slew him Cain.

26. And to Seth also to him (there) was born to him the son and he called his name Enosh. Then began (people) to call on the name of him who (is) the Lord.

www.ingramcontent.com/pod-product-compliance
Lightning Source LLC
Chambersburg PA
CBHW022118160426
43197CB00009B/1072

* 9 7 8 3 3 3 7 2 4 6 2 2 8 *